AF251296

VIBRATIONAL SPECTROSCOPY
Methods and Applications

VIBRATIONAL SPECTROSCOPY
Methods and Applications

A. FADINI, Diplom-Physiker, Dr.rer.nat.
Wissenschaftlicher Mitarbeiter Institute ITV
West Germany

F.-M. SCHNEPEL, Diplom-Chemiker, Dr.rer.nat.
Beiersdorf AG, West Germany

Translator:
CLAUS WIBBELMAN
University of Aberdeen

Translation Editor:
MARY MASSON
University of Aberdeen

ELLIS HORWOOD LIMITED
Publishers · Chichester

Halsted Press: a division of
JOHN WILEY & SONS
New York · Chichester · Brisbane · Toronto

This English edition first published in 1989 by
ELLIS HORWOOD LIMITED
Market Cross House, Cooper Street,
Chichester, West Sussex, PO19 1EB, England
The publisher's colophon is reproduced from James Gillison's drawing of the ancient Market Cross, Chichester.

Distributors:

Australia and New Zealand:
JACARANDA WILEY LIMITED
GPO Box 859, Brisbane, Queensland 4001, Australia

Canada:
JOHN WILEY & SONS CANADA LIMITED
22 Worcester Road, Rexdale, Ontario, Canada

Europe and Africa:
JOHN WILEY & SONS LIMITED
Baffins Lane, Chichester, West Sussex, England

North and South America and the rest of the world:
Halsted Press: a division of
JOHN WILEY & SONS
605 Third Avenue, New York, NY 10158, USA

This English edition is translated from the original German edition *Schwingungsspektroskopie*, published in 1985 by Georg Thieme Verlag, © the copyright holders.

© 1989 English Edition, Ellis Horwood Limited

British Library Cataloguing in Publication Data
Fadini, A.
Vibrational spectroscopy.
1. Vibration spectroscopy
I. Title II. Schnepel, F. M.
535.8′42

Library of Congress Card No. 89–15281

ISBN 0–7458–0035–1 (Ellis Horwood Limited)
ISBN 0–470–21493–7 (Halsted Press)

Typeset in Times by Ellis Horwood Limited
Printed in Great Britain by The Camelot Press Southampton

Table of contents

List of abbreviations

Commonly used abbreviations are listed here together with a reference to their first appearance in the text. Their numerous indices have been omitted.

A	non-degenerate mode of vibration, symmetric to principal axis of rotation (Table 5.5)
as	antisymmetric (Example 1.1 and Fig.1.5; section 7.2)
B	non-degenerate mode of vibration, anti-symmetric to a principal axis of rotation (Table 5.5)
$\mathbf{B}$	transformation matrix where $\Delta\mathbf{r} = \mathbf{B} \cdot \Delta\mathbf{x}$ (Eq. (11.1))
C_n	axis of rotation of multiplicity n (section 5.1.1)
c	scaling factor $f(XY) = c \cdot v^2/(\mu_x + \mu_y)$ (Eq. (7.1.1))
c	speed of light
dp	depolarized (section 3.3.3)
E	doubly degenerate mode of vibration (Table 5.5)
E	neutral element (section 5.1.1)
$\mathbf{E}$	unity matrix (section 9.1)
F	triply degenerate mode of vibration (Table 5.5)
F	force constants for symmetry co-ordinates S (Chapter 12)
$\mathbf{F}_z$	force-constant matrix (potential energy matrix) for:

$z = x$ Cartesian co-ordinates,

$z = r$ internal co-ordinates,

$z = S$ symmetry co-ordinates,

$z = Q$ normal co-ordinates (Chapter 10)

f	force constants for internal co-ordinates (Chapter 7) ($f_r = f_{xy} = f(XY)$ valence force constant for the chemical bond between the atoms X and Y; $f_\alpha = f_{(XYZ)}$ deformation force constant for the deformation of $\alpha = \angle XYZ$)
G	matrix element of the inverse matrix of the kinetic energy for symmetry co-ordinates (Chapter 12)
$\mathbf{G}_z$	inverse matrix of the kinetic energy for the co-ordinate system z (see also $\mathbf{F}_z$) (Chapter 11)
g	matrix element of the inverse matrix of the kinetic energy for internal co-ordinates (Table 5.5)

g   gerade (symmetric): index of the modes of vibration
I intensity (section 3.3.3)
IR   infrared
i centre of inversion (section 5.1.1)
ia   inactive
M   matrix of the atomic masses (Eq. (11.3))
m_i mass of the ith atom of a molecule or a solid
N number of atoms in a molecule (section 1.3.1)
N bond order (Chapter 20)
N/cm   Newton per centimetre (the units of the force constants) (section 7.1)
n degree of freedom (section 1.3.1)
n_i number of internal vibrations of the molecules (ions) in the unit cell per symmetry species (section 22.4.1)
p   polarized (section 3.3.3)
Ra   Raman
R' number of rotary vibrations per symmetry species (section 22.4.1)
r equilibrium distance between two chemically bonded atoms X and Y (often r_{XY})
S   matrix of the symmetry co-ordinates (Eq. (12.1))
S_n improper axis of rotation of multiplicity n (section 5.1.1)
s   symmetric vibration (Example 17:1 2; Example 1.1 and Fig. 1.5)
T number of accoustic vibrations per symmetry species (section 22.4.1)
T' number of translatory vibrations per symmetry species (section 22.4.1)
U   transformation matrix where $\mathbf{S} = \mathbf{U} \cdot \mathbf{\Delta r}$ (Eq. (12.1))
u   ungerade (anti-symmetric): index of the modes of vibration
α bond angle (Chapter 10)
γ dihedral (out-of-plane) angle (Chapter 10)
Δr   matrix of internal co-ordinates (section 11.1)
Δx   matrix of Cartesian co-ordinates (section 11.1)
Λ   matrix of eigenvalues (section 9.1)
λ_i ith eigenvalue where $\lambda_i = cv_i^2$ (Eqs. (9.3.5) and (13.20))
μ_i ith reciprocal mass where $\mu_i = m_i^{-1}$
N   matrix of frequencies or wave numbers (e.g. Eq. (14.15))
v_i ith frequency or wave number
ρ degree of polarization (section 3.3.3)
σ symmetry plane (section 5.1.1)
τ twist (torsion) angle (Chapter 10)

Part I
Introduction

High information content, mature technology and wide applicability have made vibrational spectroscopy one of the most important methods for structure determination. A variety of methods have been developed for the interpretation of infrared and Raman spectra. Since the specialized literature treats these methods often in an incomplete and not easily understandable way, it is the aim of this book to provide an ordered, systematic description, relating both techniques to their common basis and their areas of application.

This monograph is meant to be a textbook for undergraduate and postgraduate students of the natural sciences, as well as a resource for the spectroscopist. It is an attempt to give a systematic overview of the numerous applications of vibrational spectroscopy and an understandable description of some problem-oriented approaches.

The emphasis is put on the applicability of the various techniques: prerequisites, approaches, requirements and working methodology are discussed together with the information obtainable. Difficult theoretical concepts are elaborated only if they are required for the understanding of the method.

The structure of this book is determined by the prerequisites demanded by different applications: simple, often routinely used, techniques, which require no special knowledge and no mathematical background (Chapters 2–4), are followed by the methods utilizing group theory (Chapter 5 and 6) and the theoretically and mathematically more difficult interpretation of the vibrational spectra of free molecules and solids (Chapters 7–24).

Numerous examples of the analysis of infrared and Raman spectra and extensive tables of spectroscopic data have been incorporated in this text. These should enable the reader to apply the techniques to the solution of real problems.

1

Theoretical background

This chapter summarizes some basic concepts and definitions which are necessary for the understanding of vibrational spectroscopy; these are:

— the parameters describing electromagnetic radiation
— the number and properties of molecular vibrations and their excitation in infrared and Raman spectroscopy
— their physical basis, major differences and potential applications.

1.1 ELECTROMAGNETIC RADIATION

Electromagnetic radiation can be characterized by two quantities according to its wave nature:

— the *wavelength* λ, defined by the distance between two points on the wave of the same phase, usually given in units of m, μm or nm (1 nm = 10^{-3} mm = 10^{-9} m) and
— the *frequency* v', which is defined as the number of oscillations per unit of time, in units of sec^{-1} or hertz (Hz); 1 Hz = 1 sec^{-1}.

The product of wavelength and frequency is the speed of light c, which is 2.99793 × 10^8 m/sec in vacuum. An additional parameter is often used in vibrational spectroscopy:

— the *wave number* v, which is defined as the reciprocal of the wavelength, and has dimensions 1/length. To characterize molecular vibrations, v is best given in units of cm^{-1}.

The following relation exists between the wavelength λ, the wavenumber v, the frequency v', and the speed of light c (dimensions in [ ]):

$$v\,[\mathrm{cm}^{-1}] = 10^{-4}/\lambda\,[\mu\mathrm{m}] = v'\,[\mathrm{s}^{-1}]/c\,[\mathrm{cm/sec}] \qquad (1.1.1)$$

The energy E of the electromagnetic radiation is proportional to its frequency and its wavenumber (the reciprocal of its wavelength):

$$E = h\nu' = hc\nu = hc/\lambda \tag{1.1.2}$$

Where Planck's constant $h = 6.626 \times 10^{-34}$ J·sec.

The interaction of electromagnetic radiation with molecules is the basis of several methods for structure determination. Among the most important techniques are those which utilize the absorption (molecular spectroscopy, i.e. infrared spectroscopy), the scattering (Raman spectroscopy) or the diffraction (X-ray diffraction) of radiation. Fig. 1.1 shows the electromagnetic spectrum divided into a number of

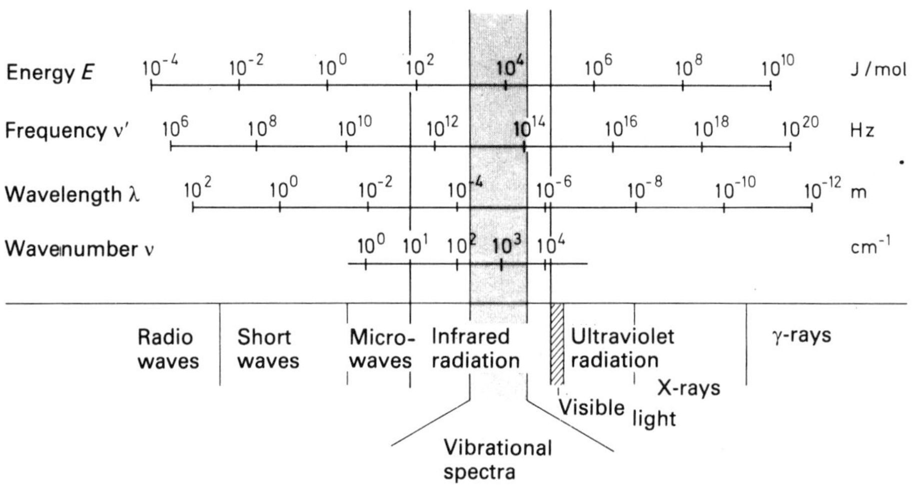

Fig. 1 — The electromagnetic spectrum.

regions: the infrared region extends from 0.8 to 1000 μm subdivided into near (0.8–2.5 μm), mid (2.5–50 μm) and far (50–1000 mm) IR. Molecular vibrations are normally observed in the mid IR, i.e. at wavenumbers from 4000 to 200 cm^{-1}.

1.2 INFRARED AND RAMAN SPECTROSCOPY

The vibrational frequencies of a molecule can be determined by two fundamentally different spectroscopic techniques:

— by using infrared spectroscopy, first systematically used by W. Coblentz in 1905 [1], and
— using Raman spectroscopy, which was theoretically predicted in 1923 by A.

Smekal [2] and experimentally observed in 1928 by C. V. Raman and K. S. Krishnan [3].

Both techniques depend on the interaction of electromagnetic radiation with matter, but the physical causes are fundamentally different. In many molecules it is therefore possible to observe vibrations of different symmetry by different spectroscopic techniques. The complete picture is often only obtained by combining the two techniques. The most important differences between infrared (IR) and Raman spectroscopy are summarized in Table 1.1.

Table 1.1 — A comparison of infrared and Raman spectroscopy

Parameter	IR spectroscopy	Raman spectroscopy
Interaction	Absorption	Scattering
Excitation of vibrations by	Polychromatic IR radiation	Monochromatic radiation (v_o) usually in the visible range
Frequency measurement	Absolute	Relative to the excitation frequency v_o
Requirement for the activity of a vibration	Change of the dipole moment $(\partial\mu/\partial Q) \neq 0$	Change of the polarizability $(\partial\alpha/\partial Q) \neq 0$
Band intensity	$I \propto (\partial\mu/\partial Q)^2$	$I \propto (\partial\alpha/\partial Q)^2$
Representation of the spectrum	Absorption logarithmic 'downwards'	Intensity linear 'upwards'
Preferred technique for	Routine analysis, gas analysis	Investigation of aqueous solutions, single crystals, polymers

1.2.1 The theory of molecular vibrations

A diatomic molecule can be represented in a macroscopic model by two masses m_1 and m_2, connected by an elastic spring (Fig. 1.2). This system will oscillate about the equilibrium distance r_e if it is first distorted by a distance Δr and then released. The oscillation is caused by the restoring force K. In the first approximation K is proportional to and opposed to the distortion (Hooke's law):

$$K = -f\Delta r = -f(r_{max} - r_e) \tag{1.2.1}$$

The proportionality factor f, which in the model corresponds to the spring constant, is called the force constant in the case of molecular vibrations. The calculation of f, which is a measure of the bond strength, is one of the most important theoretical applications of vibrational spectroscopy.

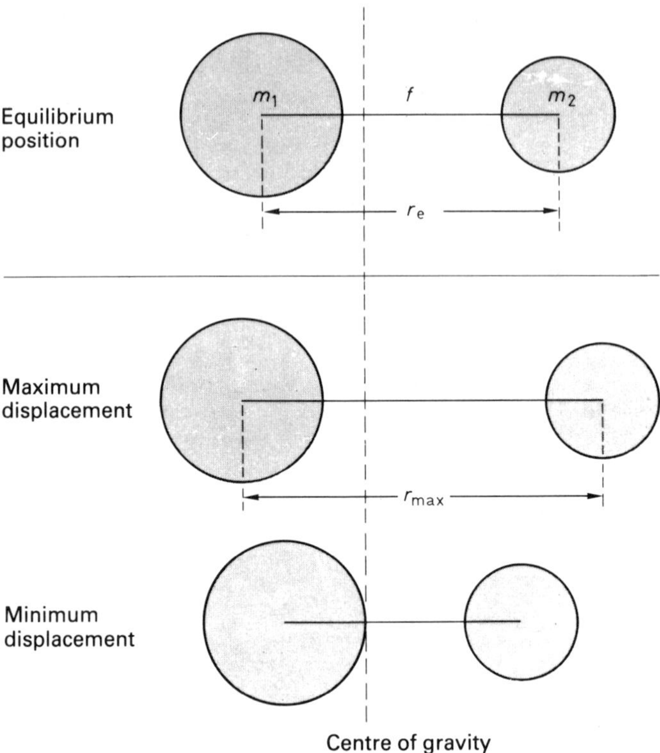

Fig. 1.2 — Model of an oscillating diatomic molecule (m_1 and m_2 atomic masses; f spring constant; r_e equilibrium separation; r_{max} maximum separation).

The harmonic case

The potential energy as a function of the interatomic distance r is obtained by integration as

$$V(r) = \tfrac{1}{2}[f(\Delta r)^2] = \tfrac{1}{2}[f(r - r_e)^2] \tag{1.2.2}$$

and has therefore a parabolic shape (Fig. 1.3(a)). The potential V and the vibrational motion are said to be *harmonic*, since they depend on a force linearly proportional to displacement (Eq. (1.2.1)). From the quantum theoretical treatment of the 'harmonic oscillator', the energy eigenvalues E_v are obtained:

$$E_v = h\nu'_{vib}(v + \tfrac{1}{2}) \qquad v = 0, 1, 2, \ldots \tag{1.2.3}$$

where ν'_{vib} is the frequency of the oscillation and v is the vibrational quantum number, which distinguishes the vibrational levels. The ground state corresponds to $v = 0$ with an energy $E = 1/2\, h\nu'_{vib}$; $v = 1$ describes the first excited level ($E_1 = 3/2\, h\nu'_{vib}$) and so on. The difference between two successive energy levels ΔE is the same for all levels ($\Delta E = h\nu'_{vib}$).

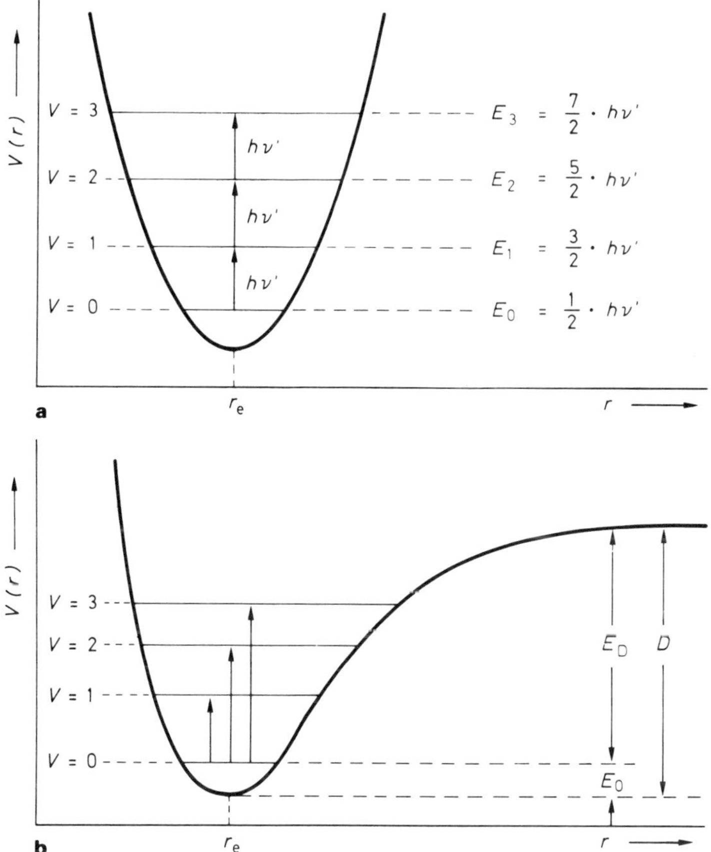

Fig. 1.3 — Potential energy curves for (a) the harmonic, and (b) the anharmonic oscillator. (E_o zero point energy; E_D dissociation energy; v vibrational quantum number).

The anharmonic case

Some phenomena, such as the dissociation of a molecule if a sufficient amount of energy is supplied, or the appearance of combination and overtones, cannot be explained with the harmonic approach. A more realistic treatment uses the anharmonic oscillator model. In this case the potential energy is approximated by the so-called Morse potential:

$$V(r) = D[1 - \exp\{-a(r - r_e)\}]^2 \tag{1.2.4}$$

This corresponds to an asymmetric potential (Fig. 1.3(b)) with curvature characterized by the constant a. D is the sum of the zero point energy E_0 and the dissociation energy E_D.

As in Eq. (1.2.3) energy eigenvalues are obtained by the quantum theoretical treatment of the anharmonic oscillator.

$$E_v = h\nu'_{vib}\,[(v + 1/2) - x(v + 1/2)^2]\quad v = 0,1,2,\ldots \tag{1.2.5}$$

The anharmonicity constant x is a measure of the deviation from the harmonic case. In real systems $x > 0$, so the energy difference between successive vibrational levels decreases as v increases until the dissociation energy of the molecule is reached.

1.2.2   IR spectroscopy

The excitation of a *fundamental vibration* can be described as the transition from the vibrational ground state to the next energy level on absorption of one quantum of light (cf. also Table 1.2). This transition is only possible if the energy difference between the two vibrational levels is equal to the energy of the photon (resonance condition):

$$E_1 - E_0 = h\nu'_{vib} = E_{ph} = h\nu'_{ph} \tag{1.2.6}$$

Thus the basic requirement is that $\nu'_{vib} = \nu'_{ph}$ (ph = photon).

The infrared spectrum of a sample is measured by exposing it to *polychromatic infrared radiation*. The vibrational frequencies are obtained as the absolute frequencies of the absorbed radiation by comparing the intensity of the sample beam with a suitable reference beam at all wavelengths. Fundamental transitions (according to the selection rule $\Delta v = +1$) and, owing to anharmonicity, also those with $\Delta v = +2$, $+3$, etc., are allowed. However, the transition probability and hence the intensity of the respective absorption bands decreases with increasing energy difference between the initial and final states. The transition $v = 0$ to $v = 1$ corresponds to the fundamental vibration and $v = 0$ to $v = 2$ to the first overtone. This results in a considerably weaker absorption band at slightly less than double the fundamental frequency.

1.2.3   Raman spectroscopy

When a substance is irradiated with intense light of frequency ν'_0, which it cannot absorb, a small part of the incident radiation is scattered in all directions, and ν'_0 additional frequencies are observed in the spectrum of the scattered light (Raman spectrum). The connection between these additional frequencies and the vibrational frequencies of the molecule can be illustrated in various ways:

— From the wave theory of light one can postulate forced vibrations for the electrons of an atom in the electromagnetic field. The excitation frequency will be modulated by the eigenfrequencies ν'_{vib} of the molecules according to $\nu'_0 \pm \nu'_{vib}$.

— In the quantum mechanical model the incident light beam consists of photons with energy $E = h\nu'_0$ and the scattering process is explained in terms of collisions of these photons with molecules of the sample. In the case of elastic scattering the energy of both molecules and photons remains unchanged ($\nu' = \nu'_0$; Rayleigh scattering). Energy changes occur during inelastic collisions, and additional frequencies $\nu' \neq \nu'_0$ are observed in the scattered light. In a simplified diagram (cf. Table 1.2) three cases are discernible:

(1) A molecule is promoted by interaction with a photon ($E = h\nu'_0$) from the

Table 1.2 — Comparison of transitions in infrared and Raman spectroscopy ($v = 0$ vibrational ground state; $v = 1$ first excited state; broken lines indicate hypothetical states)

Infrared spectroscopy	Raman spectroscopy		
	Energy balance: $h\nu'_o + \frac{1}{2}mv_o^2 + E_o = h\nu'_1 + \frac{1}{2}mv_1^2 + E_1$ The energy change of the molecule $\Delta E_M = E_1 - E_o$ is well approximated by: $\Delta E_M = h(\nu'_o - \nu'_1)$		
Absorption	(a) Inelastic collision $E_1 > E_o$; $\Delta E_M > 0$	(b) Elastic collision $E_1 = E_o$; $\Delta E_M = 0$	(c) Inelastic collision $E_1 < E_o$; $\Delta E_M < 0$
$\nu'_{vib} = \nu'_{LO}$	$\nu'_2 = \nu'_o - \nu'_{vib} < \nu'_o$	$\nu'_1 = \nu'_o$	$\nu'_3 = \nu'_o + \nu'_{vib} > \nu'_o$

vibrational ground state ($v = 0$) into a much higher, unstable state and immediately drops back to the ground state. No energy changes occur and the scattered photon retains its original frequency: $v'_1 = v'_0$. This causes the most intense 'Rayleigh' line in the spectrum of the scattered radiation.

(2) The molecule does not return to its ground state, but to the first excited level ($v = 1$). In this case it retains some of the energy, and the energy of the photon after the collision will consequently be reduced, i.e. $v'_2 = v'_0 - v'_{vib}$ ('Stokes line').

(3) An interaction takes place between a photon and an excited molecule ($v = 1$) and the molecule returns to its ground state. The scattered photon will then have a higher energy $v'_3 = v'_0 + v'_{vib}$. The corresponding 'Anti-Stokes' line is normally weaker than the Stokes line since most molecules are in the ground state at room temperature.

Note that the vibrations are excited by *monochromatic radiation* which is not absorbed by the molecules. Radiation with an energy somewhere between the IR region and the region of electronic transitions is used for this purpose (e.g. a helium–neon laser: $\lambda = 632.8$ nm, $v = 15802$ cm^{-1}). Normally only the Stokes lines are recorded. The frequency scale is not absolute but *relative* to the excitation frequency: $v'_{vib} = v'_0 - v'$.

1.3 VIBRATIONS

The spatial arrangement of a molecule changes during (cf. Table 1.3)

— translations, when the molecule moves as a whole, as described by the changing co-ordinates of its centre of gravity;
— rotations about the principal axis; and
— vibrations, during which bond distances and/or angles change periodically.

Table 1.3 — The principal types of motion

Motion	Symbolic representation	Remains constant	Changes
Translation		Bond lengths and angles	Centre of gravity of the molecule
Rotation		Centre of gravity, bond lengths and angles	
Vibration		Centre of gravity of the molecule	Bond lengths and/or angles

1.3.1 Fundamental or normal modes of vibration

Fundamental or normal modes of vibration are defined by the following conditions:

— All atoms of the molecule move with the same frequency and usually in phase. That is they pass the zero crossings and turning points at the same time. The amplitudes, however, depend on the masses.
— The vibrational motion does not cause a translation or a rotation of the molecule as a whole.
— Each normal mode of vibration can be excited independently.

A normal vibration is usually not localized at a single bond in a molecule, but includes several atoms. The vibrational motion is described by a so-called 'normal co-ordinate' Q, which is derived from the masses and relative movements of the atoms involved.

The number of normal modes of vibration and consequently the number of normal co-ordinates follow from simple considerations:

A system of N particles has $3N$ degrees of freedom corresponding to the three independent co-ordinates of each of the N particles. Three of these are taken up by the translations of the entire molecule along the x-, y- and z-axes and another three by the rotation of the molecule about the three principal axes of inertia.

Linear molecules have only two rotational degrees of freedom because the moment of inertia along the molecular axis is zero.

The number of remaining vibrational degrees of freedom n is identical to the number of fundamental vibrations, that is:

$$n = 3N - 3 - 3 = 3N - 6 \text{ for a non-linear molecule and}$$
$$n = 3N - 3 - 2 = 3N - 5 \text{ for a linear molecule.}$$

It will become evident in the following that a practical description of the vibrational degrees of freedom is not based on Cartesian but on 'internal' co-ordinates, which correspond to bond lengths and angles.

Classification of normal modes of vibration

Various criteria can be used to characterize the ($3N$–6) or ($3N$–5) normal modes of vibration (Table 1.4 and Fig. 1.4):

(1) Classification by type.
  Four main types of vibration can be distinguished:
  (a) During valence or stretching vibrations (symbol v) one or more of the bond lengths change.
  (b) Planar bending vibrations (symbol δ); one or more bond angles change, while bond lengths remain constant.
  (c) Out-of-plane bending vibrations (symbol γ); one atom oscillates through a plane defined by (at least) 3 neighbouring atoms.
  (d) Torsion vibrations τ; a dihedral angle (the angle between two planes, which have one bond in common) is changed.

The frequencies of these four types of vibration decrease in the order: $v > \delta > \gamma > \tau$.
  Special types of vibration, which can occur in larger molecules, can be derived

Table 1.4 — The basic types of vibration

Type	Stretching vibrations	Bending vibrations		Torsion vibrations
		In-plane	Out-of-plane	
Schematic				
Symbol	ν	δ	γ	τ
minimum number of bonds	1	2	3	3

from these basic types. In organic molecules, for example, the bending vibrations of the CH_3 group are further classified into rocking ρ, twist t, and wagging ω vibrations (Fig. 1.4).

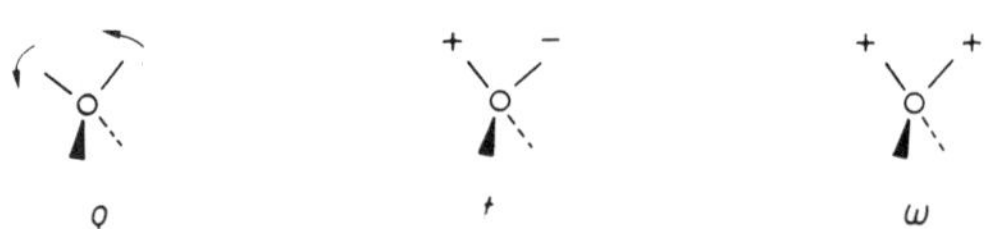

Fig. 1.4 — Special types of vibration: ρ rocking; t twist; ω wagging.

(2) Classification by symmetry.

There are

(a) symmetric vibrations (index s), where the symmetry of the molecule is retained throughout, and
(b) asymmetric vibrations (index as), where one or more of the symmetry elements of the molecules vanish during the vibration.

In highly symmetric molecules
(c) degenerate vibrations can occur. In this case two or more vibrations, depending on the degree of degeneracy, have different co-ordinates but the same energy. They occur therefore at the same position in the spectrum.

Example 1.1   Fundamental vibrations of triatomic molecules
A comparison of the normal vibrations of carbon monoxide (CO_2) and sulphur dioxide (SO_2) illustrates the various vibrations of linear and non-linear molecules (Fig. 1.5): both molecules have two stretching vibrations (v) — one symmetric v_s where both bonds are stretched simultaneously, and one asymmetric v_{as} where one bond is compressed while the other is elongated. While the SO_2 molecule (Fig. 1.5(a)) has only one more vibration δ_s, the CO_2 molecule, as a consequence of its linearity, has two bending vibrations (Fig. 1.5(b)). One occurs in the plane of the paper, the second is obtained by a 90° rotation about the molecular axis. Both have the same energy: they are degenerate (δ_c). Any attempt to construct a second bending of the SO_2 molecule would result in a rotation (Fig. 1.5(b)).

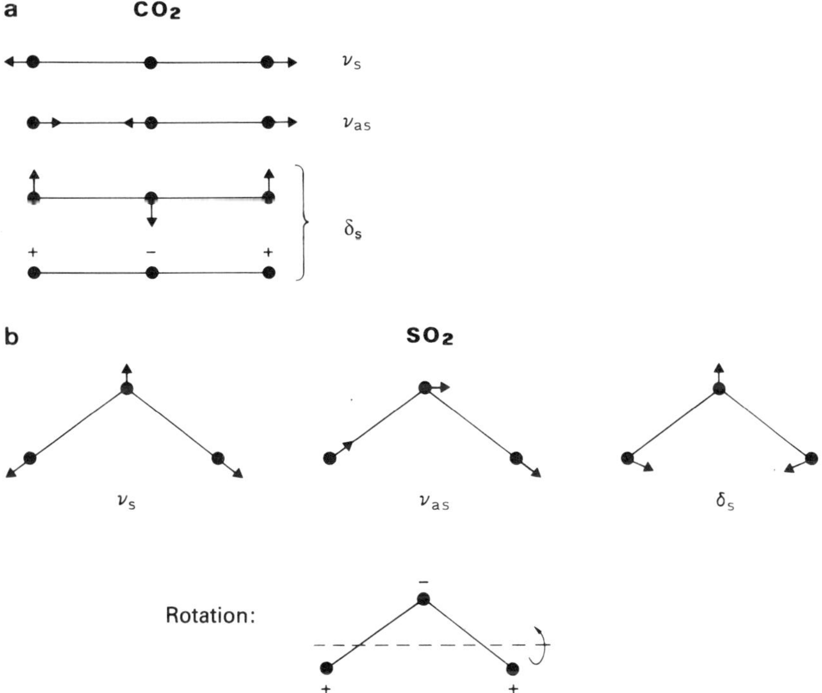

Fig. 1.5 — Normal vibrations of triatomic molecules: (a) linear molecule (CO_2); $3 \times 3 - 5 = 4$ normal vibrations; (b) non-linear molecule (SO_2); $3 \times 3 - 6 = 3$ normal vibrations.

1.3.2   Combinations and overtones
Two further types of vibration have to be distinguished from the normal or fundamental vibrations:

— combination bands occur at frequencies v'_c and are obtained by the combination of the frequencies (or multiples thereof) of one or more fundamental vibrations ($v'_c \approx a v'_1 \pm b v'_2 \pm \ldots$ where a, b are integers);

— overtones occur at approximately twice or three times the frequencies of the respective fundamental vibrations ($v'_{0,n} \simeq nv'_1$).

Group theory and the symmetry of the molecule determine which of these vibrations are 'allowed' and observed in the spectrum (see section 5.6).

Combinations and overtones are usually observed with much lower intensities than normal vibrations.

The Fermi resonance represents a special case: if a combination or an overtone coincides approximately with a fundamental vibration then the frequencies move apart and two separate bands with more or less equal intensities are observed. The separate assignment of these bands to different vibrations is then impossible.

Example 1.2   *Propionaldehyde*
The first overtone of the $H-C=O$ deformation has almost the same frequency as the $C-H$ stretching vibration of the aldehyde group. In this case, as a result of Fermi resonance, two bands of similar intensity at 2728 and 2830 cm^{-1} are observed instead of a strong fundamental and a weak overtone band. This feature can be used for the identification of aldehydes.

1.3.3   Infrared and Raman activity of vibrations

Not every one of the $3N–6$ or $3N–5$ possible normal vibrations will be observable as a band in the IR or Raman spectrum. There are generally some vibrations of symmetric molecules which are inactive and therefore not observed in either of the spectra (seldom in both).

Basic rules
The following rules apply generally:

— A vibration is IR-active only if it alters the dipole moment μ of the molecule:

$$(\partial\mu/\partial Q) \neq 0 \tag{1.3.1}$$

Q is the normal co-ordinate of the vibration (section 1.3.1). It is unimportant whether the molecule has a permanent dipole moment or not. Only the change associated with the vibration is important.

— In analogy to this, a vibration is Raman-active only if it modifies the polarizability α of the molecule (section 1.3.3):

$$(\partial\alpha/\partial Q) \neq 0 \tag{1.3.2}$$

These rules are given here without proof. A theoretical treatment can be found in [4] and [5].

Detailed analysis of vibrational spectra usually requires some knowledge of the number and activity of the vibrations. In a practical approach these are not determined by geometrical considerations, but from group theoretical arguments. These will be discussed in detail in Chapters 5 and 6. It should be obvious that infrared and Raman spectroscopy do not always show the same vibrations, but that the two spectra are complementary.

An illustration

A molecule has a *dipole moment* if the centre of charge of the electron density distribution is spatially separated from the positive charge of the nucleus. $\mu = el$ for two point charges $q_1 = -e$ and $q_2 = +e$ separated by the distance l. The dipole moment can be represented graphically by an arrow from the negative to the positive charge (Fig. 1.6).

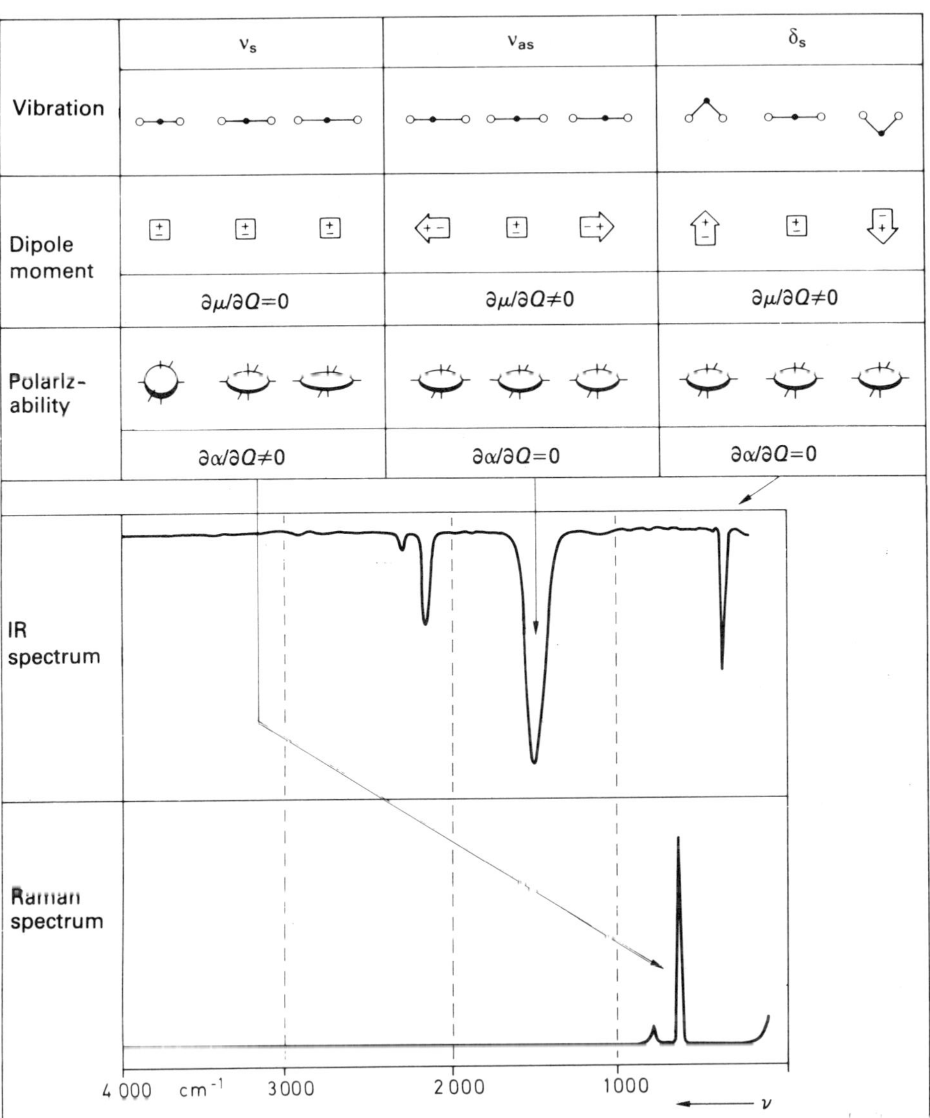

Fig. 1.6 — The changes of dipole moment and polarizability during the normal vibrations of the carbon disulphide molecule (S=C=S) and the resulting appearance of the IR and Raman spectra.

Polarizability. An electromagnetic field will distort the position of the electron cloud of a molecule relative to the nuclei. This induces a dipole moment **M**, which is proportional to the strength of the electric field **E**:

$$\mathbf{M} = \alpha\mathbf{E} \tag{1.3.3}$$

The polarizability connects the vectors **E** and **M**. α can be thought of as a measure of the 'softness' of the electron cloud.

Eq. (1.3.3) can also be written as

$$\mathbf{M}_x = (\alpha_{xx}\mathbf{E}_x) + (\alpha_{xy}\mathbf{E}_y) + (\alpha_{xz}\mathbf{E}_z)$$
$$\mathbf{M}_y = (\alpha_{yx}\mathbf{E}_x) + (\alpha_{yy}\mathbf{E}_y) + (\alpha_{yz}\mathbf{E}_z) \tag{1.3.4}$$
$$\mathbf{M}_z = (\alpha_{zx}\mathbf{E}_x) + (\alpha_{zy}\mathbf{E}_y) + (\alpha_{zz}\mathbf{E}_z)$$

The principal axes of the polarizability x',y' and z' can be defined in addition to the general co-ordinate axes x, y and z in such a way that with respect to the nine coefficients α_{ab} only $\alpha_{x'x'}$, $\alpha_{y'y'}$ and $\alpha_{z'z'}$ have non-zero values. The 'polarizability ellipsoid' is obtained by drawing a length of $1/\sqrt{\alpha}$ from the origin of these axes and connecting their end points (Fig. 1.6). A vibration is Raman-active only if it alters the size, shape or orientation of the polarizability ellipsoid.

Example 1.3   The vibrational spectra of carbon disulphide (CS_2)
Fig. 1.6 illustrates the changes of the dipole moment μ and the polarizability α of the CS_2 molecule during the normal vibrations. Both the ν_{as} and the δ_e vibrations are IR-active because they alter the dipole moment of the molecule even though it does not possess a permanent dipole moment. The polarizability ellipsoid changes only during the ν_s vibration; this is the only Raman-active vibration of the CS_2-molecule. Together the IR and Raman spectra illustrate the so-called mutual exclusion principle: all fundamental vibrations of molecules which have a centre of symmetry (section 5.1.1) are either IR- or Raman-active, but not both. Therefore the bands in the two spectra cannot coincide.

It is evident already at this stage that the structure of the molecule has a strong influence on the appearance of the spectra: its geometry determines the symmetry of the vibrations and these in turn determine changes in the dipole moment μ and polarizability α. These inter-relations will be described in more detail in the chapters dealing with group theory.

1.3.4   The number of bands in a spectrum

An N-atomic molecule can have $3N-6$ or $3N-5$ normal modes of vibration. This number, however, does not always correspond to the number of observed bands. It can be greater or less than the theoretically predicted number of bands and is determined by several factors:

An increase in the number of bands is caused by:
— the appearance of combination bands and overtones (section 1.3.2) which can be
  complicated further by Fermi resonance, and
— the existence of different configurations of the molecule in the sample. This of
  course only increases the number of bands observed in the spectrum, but not the
  number of bands of any of the species, which have to be dealt with separately (cf.
  Example 6.4)

A decrease in the number of bands is caused by:
— the degeneracy of vibrations (two or more vibrations of the same structural unit
  of the molecule have the same energy — section 1.3.1)
— similar vibrations of corresponding, but spatially separate, parts of the molecule
  (for example the vibrational frequencies of several CH_3-groups in the same
  molecule often coincide);
— accidental degeneracy (two or more unrelated vibrations have the same
  frequency);
— inactivity of vibrations (section 1.3.3).

The number of bands can also be reduced because of instrumental factors, i.e.
insufficient resolution of the spectrometer or inaccessibility of low wavenumbers.

1.3.5 Summary

Infrared and Raman spectroscopy, like other spectroscopic methods, are based on
the interaction of electromagnetic radiation with matter. The aim is the determination of the vibrational frequencies of molecules. In many cases only the combination of IR and Raman spectra will give a complete picture, because of their
different physical bases.

The fundamental task in all spectroscopic investigations is the interpretation of
the measured frequencies. The demands on the information content depend on the
problem. The interpretation may therefore require different approaches (cf. Table
1.5). The methods discussed in the following chapters are grouped into sections
according to type and degree of difficulty.

Table 1.5 — Some applications of vibrational spectroscopy and their prerequisites

Prerequisites	Resources	Applications, information obtainable	Chapters
IR or Raman spectrum, no special knowledge	Spectra for comparison, libraries of spectra	Identity testing — materials testing — process monitoring, e.g. distillation, etc.	2, see p. 33
IR and/or Raman spectrum, partial assignments of some (characteristic) bands, reference spectra	As above, plus tables for the empirical correlation of molecular structure and vibrational spectra	Identification of unknown substances and functional groups — qualitative analysis — comparison of homologues — absorption mechanisms — quantitative analysis	3, see p. 40 4, see p. 63
Complete IR and Raman spectra assignment of all bands, chemical analysis	As above, plus group theory	Determination of molecular symmetry — distinguishing different modifications — investigation of mixtures of isomers	5–6, see p. 73
Complete IR and Raman spectra, assignment of all bands, structural data	As above, plus mathematical techniques and computer programs	Force constants, assignments of vibrations, estimates of bond lengths and bond types	7–16, see p. 109 17, see p. 163 19, see p. 170 20, see p. 172

Part II
Non-mathematical interpretation of vibrational spectra

2

Applications not requiring special knowledge

Vibrational spectroscopy can be applied to some problems without any theoretical background or even basic knowledge of spectral interpretation. Generally these are tasks which require only the comparison of a spectrum with existing material.

This chapter will give two simple examples taken from different areas of application:

— the determination of whether one substance is identical to another (fingerprint spectra, identity testing; section 2.1);
— the monitoring of a reaction by recording spectra continuously (section 2.2).

2.1 IDENTITY TESTING (FINGERPRINTING)

Spectra used for this purpose do not have to meet any special requirements, except that they should be obtainable rapidly and routinely. Because of the simpler instrumentation, infrared spectra are usually preferred for these tasks, except for the cases when special experimental conditions apply, e.g. for the study of aqueous solution (Example 2.2).

2.1.1 Comparison of spectra

Some points should be noted:

The sample and reference substances should be

— in the same physical state. The vibrational frequencies of gaseous compounds are often higher than those of the same molecule in the liquid state (differences are approx. 1% [6]). For solids, these changes are even bigger. The vibrational spectra of solids can be further complicated by bands which do not occur in the spectra of of the gaseous or liquid sample (removal of the inactivity of some vibrations, section 6.1.1);
— measured by using the same techniques and conditions. To obtain IR spectra from a solid sample, it could be dissolved in a suitable solvent, dispersed in a mull

(Nujol, KelF) between windows or dispersed with another solid (KBr, CsI) and formed into pellets. Physical and/or chemical interactions of the sample with the solvent or dispersing matrix can alter the spectrum to such an extent that proof of identity is no longer possible.

— checked for correspondence of all bands:

The bands in a vibrational spectrum can be classified into

(1) Fundamental vibrations, combinations and overtones

In section 1.3.1 it was shown that a maximum of $3N-5$ or $3N-6$ bands belong to the fundamental vibrations of an N-atomic molecule. All other bands can be derived from these by combination. This classification is important for the assignment since in most cases the strongest bands belong to the fundamental transitions (exception Fermi resonance, Example 1.2);

(2) Characteristic vibrations

A number of bands result from vibrations, which are characteristic of certain structural elements (section 3.1). These bands occur in the spectra of all molecules which contain these groups. They are important for the classification of unknown compounds, but not suitable for identity testing. Vibrations which involve large parts of the molecule are more important in the present context. They are observed at wave numbers below 1200 cm^{-1} and are not well suited for diagnostic purposes. However, they are the most characteristic feature in the spectrum of a substance, comparable to the fingerprint used in police work.

Bands in this region of the spectrum are not characteristic of the class of compounds but of the compound itself. It is therefore called the "fingerprint region" of the spectrum (Example 3.4).

These classifications are of fundamental importance for the applications described in the following chapters, but they are of minor importance for the present task of comparing spectra:

The comparison of spectra requires that the position and the intensities of all bands (not only the most intense) are considered (point-to-point comparison, such as Example 3.4).

The rules also apply to mixtures of compounds. A quantitative or semi-quantitative analysis of the spectra (Chapter 4) is required to permit conclusions to be reached about changes of concentration or composition.

2.1.2 Applications

The question of whether one substance is identical to another is considered in the following examples. Vibrational spectroscopy is chosen to solve this problem because the IR and Raman spectra are characteristic of a substance (as are other spectra and physical data).

Vibrational spectroscopy is preferred for the investigation of pure substances over other physical techniques which yield only one parameter (boiling point, refractive index). Infrared spectroscopy is often preferred to other spectroscopic techniques because of the relative simplicity of the instrumentation.

An identity check may be required for several reasons:

— To find out whether substances which have been produced or delivered at different times are identical. Comparison of the IR spectra with those of authentic samples allows simple and fast quality and process control.
— To determine the stability of a compound over a period of time, or test whether decompositions or rearrangements occur (Example 4.2, nitrito $\rightarrow$ nitro isomerization).
— To verify that a synthetic compound is identical to a natural product. The comparison of spectra can often replace a series of physical measurements (melting, boiling point, refractive index, etc.).
— To investigate whether a compound retains its structure in different phases or physical states. This can involve the gaseous, liquid and solid states, and also solutions, etc. Slight band shifts have to be allowed for. The investigation of aqueous solutions by IR spectroscopy is often impossible or at least very difficult because of the broad, very intense bands of the solvent and the water solubility of the commonly used infrared transparent window materials. In these cases the use of Raman spectroscopy is advantageous.

Example 2.1   Isopolymolybdates in the solid state and in aqueous solution [7]
The spectra of solid molybdates of the general formula $A_8Mo_{36}O_{112}.$ *ca.* 80 H_2O

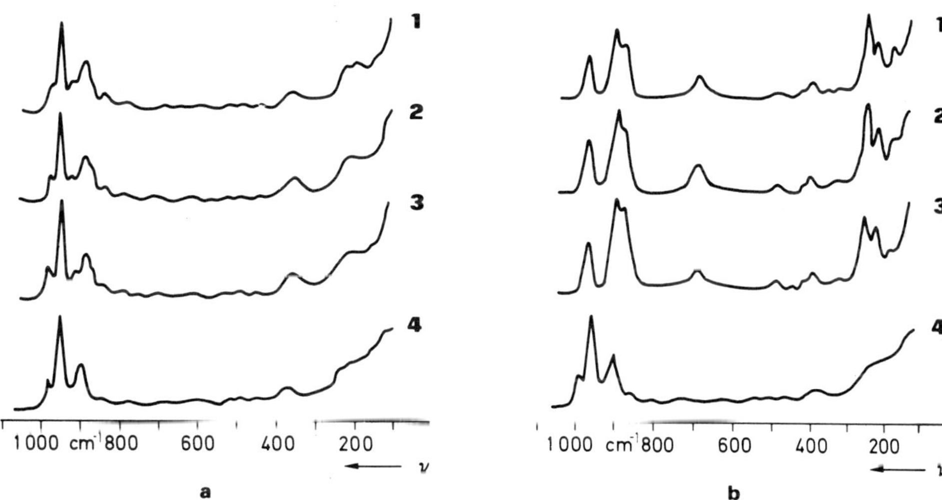

Fig. 2.1 — Raman spectra of polymolybdates: (a) $A_8Mo_{36}O_{112} \cdot$ *ca.* 80 H_2O; (b) (AHMo$_5$ O$_{16}$(H$_2$O)H$_2$O$_n$ spectra of the solids $A^+ = NH_4^+$ (**1**), K^+ (**2**), Na^+ (**3**); spectrum of the aqueous solutions (**4**) (reprinted from K. H. Tykto and B. Schönfeld, *Z. Naturforsch.*, 1975, **30b**, 471 by permission of the copyright holders).

$(A^+ = Na^+, K^+, NH_4^+)$, which consist of highly aggregated discrete polyanions, are identical to spectra obtained from aqueous solutions. Obviously the same species exist in both phases. The solid and solution Raman spectra of polymolybdates $(AHMo_5O_{16}(H_2O)H_2O)_n$, which in contrast consist of chains of octahedra, differ significantly. In this case the presence of different polymolybdate species in the two phases (Fig. 2.1) seems likely.

2.1.3 Testing for impurities

As the examples have demonstrated, the identity testing of pure substances is relatively simple and fast, and in most cases vibrational spectra can provide strong evidence. However, conclusions cannot necessarily be drawn about the purity of a substance. The existence of impurities can only be ascertained if they produce distinguishable new features in the spectrum. This depends on several factors:

— the intensity of the band(s) of the contaminant and the similarity of its spectrum to that of the main component and
— the parameters of the measurement, i.e. optical path length in the sample and facilities for rescaling the spectra.

Example 2.2   Acetone in methanol
To illustrate these points, we have chosen a compound (acetone) with a very strong band ($v = 1712$ cm^{-1}). Fig 2.2 shows the IR spectra of methanol (**1**) and

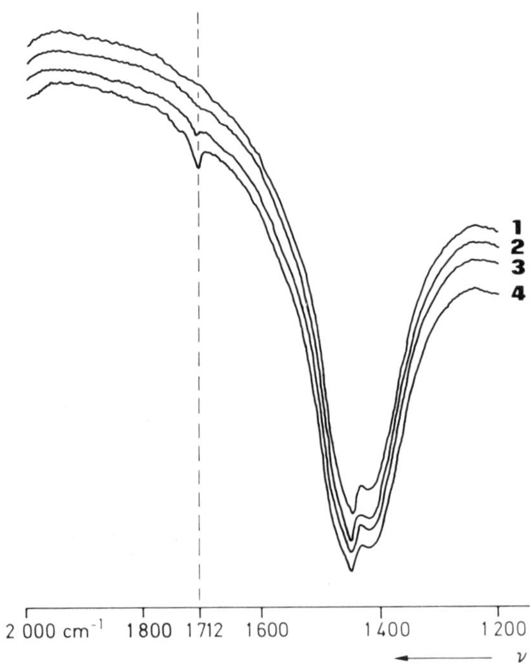

Fig. 2.2—Segment of the IR spectrum of pure methanol (**1**) and mixtures with 0.05% (**2**), 0.1% (**3**) and 0.2% (**4**) w/w of acetone.

mixtures containing 0.05 (**2**), 0.1 (**3**) and 0.2 (**4**) % w/w of acetone. On an ordinate scale length l of 20 cm, 0.1% of acetone (**3**) is clearly distinguishable, but 0.05% (**2**) is barely recognizable.

Generally the lower detection limit for impurities is 0.5% by weight, but in special cases it can be as low as 0.03%. In unfavourable circumstances it is much higher, for example when weak bands have to be used or when the spectra are very similar. This is the case for n-octane in n-decane, where the detection limit is 10% for an optical path of 0.01 cm [8]. It is obvious from these values that vibrational spectroscopy is relatively insensitive to traces of impurities. When very stringent requirements for purity have to be met, other techniques are preferred, such as gas chromatography.

2.2 FOLLOWING THE COURSE OF REACTIONS

Reactions with low rate constants can be monitored by continuous recording of vibrational spectra. The time dependence of the band intensities of the starting materials, intermediates and products can be used to control the reaction.

Example 2.3   Preparation of N-pentafluoroethyltetrafluorosulphurimide
        $(C_2F_5-N=SF_2)$ *[9]*

The following reactions occur during the fluorination of N-pentafluoroethyldifluorosulphurimide ($C_2F_5-N=SF_2$, **1**), under ultraviolet irradiation, in Pyrex reaction vessels:

$$C_2F_5-N=SF_2 \quad \xrightarrow{F_2/\,UV}$$
1

$$C_2F_5-N=SF_4 \quad \xrightarrow{F_2/\,UV} \quad C_2F_5-NF-SF_5$$
2   **3**

$$C_2F_5-N=SF_2=N-C_2F_5 \quad \xrightarrow{F_2/\,UV}$$
4

$$C_2F_5-NF-SF_4-NF-C_2F_5$$
5

The primary products **2** and **4** react slowly with fluorine. It is possible to isolate **2**, which is only with difficulty obtained by other methods, if the reaction is terminated at the right moment. This can be achieved by connecting a gas cell to the reaction flask in a way that allows the continuous sampling of IR spectra (Fig. 2.3(a)). The spectra of all compounds that take part in the reaction are required for reference purposes (Fig. 2.3(b)–(f)) It is possible to select at least one characteristic band for each compound, even though their spectra are very similar. The spectra are recorded at 3–5 min intervals, and the following spectral changes are observed:

Apparatus

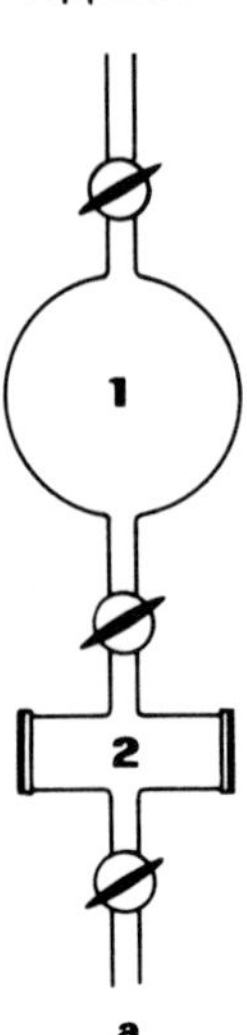

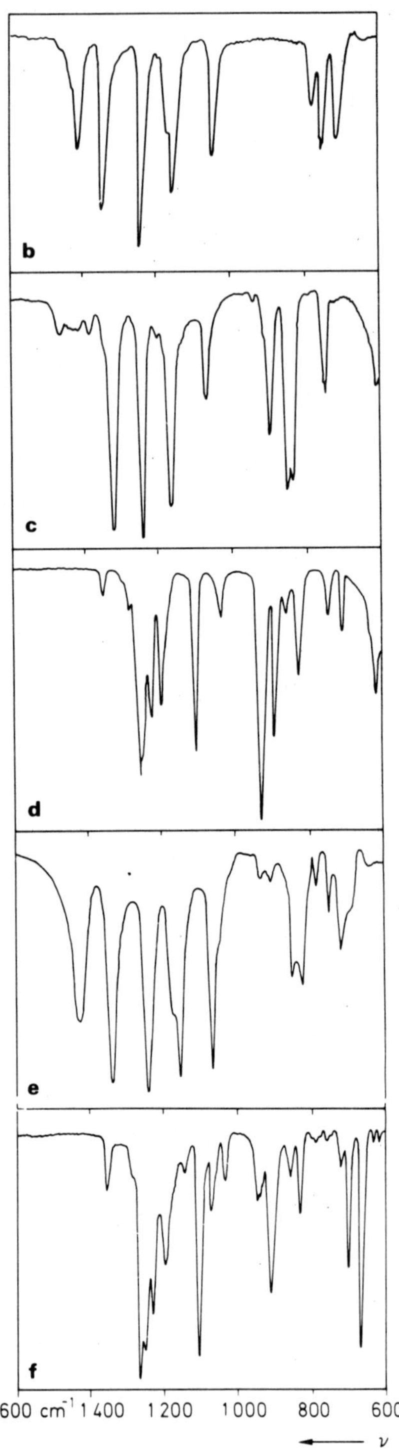

Fig 2.3 — (a) Apparatus for continuous sampling and measurement of IR spectra (**1** — reservoir, **2** — IR gas cell); (b) $C_2F_5-N=SF_2$; (c) $C_2F_5-B=SF_4$; (d) $C_2F_5-BF-SF_5$; (e) $C_2F_5-N=SF_2=N-C_2F_5$; (f) $C_2F_5-NF-SF_4-NF-C_2F_5$.

The band intensities of the starting material N-pentafluoroethyldifluorosulphurimide ($C_2F_5-N=SF_2$, **1**, Fig. 2.3(b)) decrease, and new bands, which indicate the formation of the intermediates, appear simultaneously:

$C_2F_5-N=SF_4$ (**2**) at wavenumbers $v = 1315$ and 1066 cm^{-1} (Fig. 2.3(c)) and $C_2F_5-N=SF_2=N-C_2F_5$ (**4**) at $v = 1429$ and 1066 cm^{-1} (Fig. 2.3(e)).

The next reaction step is observed approximately 30 min after the start of the reaction, when the band at 1315 cm^{-1} weakens and those belonging to $C_2F_5-NF-SF$ (**3**) at $v = 1220$, 1193 and 1032 cm^{-1} (Fig. 2.3(d)) appear in the spectrum. The best yield of $C_2F_5-N=SF$ (**2**) is obtained by stopping the irradiation at this point and isolating the product by chemical or physical means.

In a similar application, band intensities can be monitored to determine reaction kinetics (Example 4.2, p. 69).

Example 2.3 is already slightly more involved than Example 2.1 insofar as it requires the assignment

$$\text{band} \rightarrow \text{molecule}$$

This is the first and simplest assignment which can be made. It is only useful for simple systems where at least one band is significantly different for each molecule. The assignment

$$\text{band} \rightarrow \text{structure element of the molecule}$$

is the basis for the classification of compounds. It is the first stage towards the identification of a compound and requires additional models and tools. This is especially true for the assignment

$$\text{band} \rightarrow \text{vibration of the molecule,}$$

which stands at the beginning of every thorough spectroscopic analysis.

The theory basis and applications of these methods for assignment will be described in the chapters that follow.

3

Applications requiring some background in spectroscopy

Any application that goes further than the mere comparison of spectra requires at least a partial assignment of the vibrations of a molecule. The first two parts of this chapter deal with some important applications which require the analysis of some regions of a spectrum. The interpretation is often limited to a few frequencies. It is based on the assignment band $\rightarrow$ structure element and is mainly concerned with the position of the most intense bands.

The criteria for the more exact assignment band $\rightarrow$ molecular vibration will be described in the third part of this chapter. They are the basis for all later sections and require the interpretation of further spectroscopic parameters, and some theoretical knowledge.

3.1 IDENTIFICATION OF UNKNOWN COMPOUNDS

The classification and subsequent identification of an unknown compound is one of the most frequent applications of vibrational spectroscopy. The main advantage over other techniques is the fact that no mathematical knowledge of the basic principles of vibrational spectroscopy is presumed and only a comparison with empirically compiled material is required.

The most frequently used method involves the interpretation of regions of characteristic frequencies.

3.1.1 Characteristic spectral regions

Classical mechanics describes the relationship between the vibrational frequency v'_{vib} of a diatomic molecule, the bond strength (expressed by the force constant f) and the atomic masses m_1 and m_2 as follows:

$$v'_{vib} = 1/(2\pi)\sqrt{f(m_1 + m_2)/(m_1 m_2)(1 - 2x)} \tag{3.1.1}$$

The factor $(1-2x)$ accounts for the anharmonicity of the vibration (section 1.2.1). It can usually be neglected.

The relationship between v'_{vib}, f and m allows a first coarse partitioning of the spectrum (Fig. 3.1):

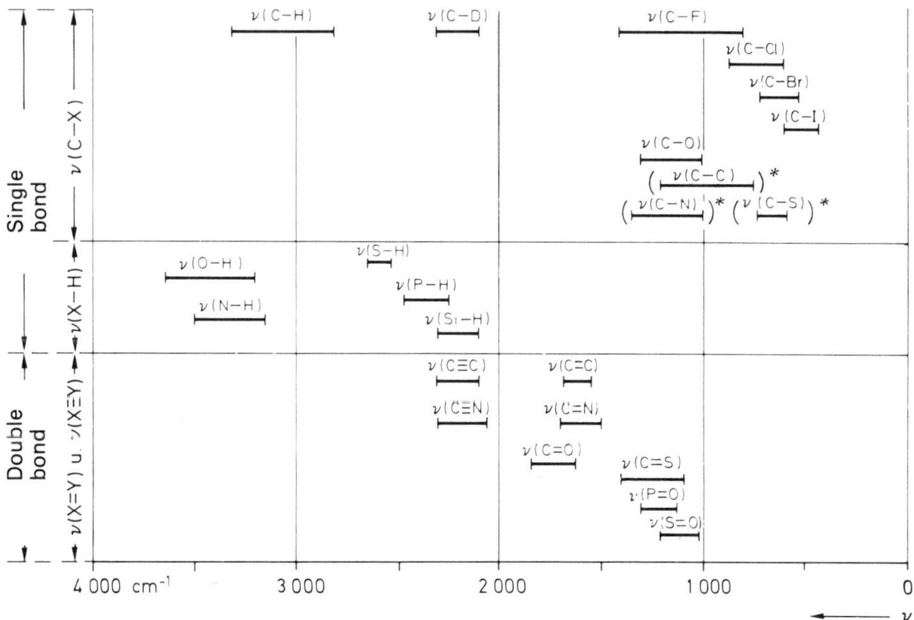

Fig. 3.1 — Partitioning of the vibrational spectrum into regions which are characteristic of some stretching vibrations (*marks regions which are less characteristic).

— The vibrational frequency increases with increasing bond strength ($v'_{vib} \propto \sqrt{f}$), for example

$$v'(C \equiv C) > v(C = C) > v'(C - C).$$

— The vibrational frequency decreases when the relevant atomic masses increase:

$$v'(C-H) > v'(C-O) > v'(C-Cl), \text{ etc.}$$

These rules are limited in application, and they may not be valid if strong vibrational coupling occurs.

Group frequencies
Some bonds or functional groups can be treated as independent oscillators even if they are part of a larger molecule. Characteristic normal vibrations are then localized

to that structure element and can be described entirely by the motions of the relevant atoms†. These vibrations are largely independent of the rest of the molecule so that their frequencies appear in a relatively narrow region of the spectrum.

Example 3.1 Vibrational spectra of monosubstituted ethylene derivatives (Table 3.1)

The IR spectra of compounds containing $-CH=CH_2$ always exhibit a band in a narrow spectral region.

Table 3.1 — Wavenumbers of monosubstituted ethylene derivatives (reprinted from [10] by permission of the copyright holders, Verlag Chemie)

Compound	v [cm^{-1}]
$CH_3-CH=CH_2$	1651
$CH_3)_3C-CH=CH_2$	1641
$CH_3-(CH_2)_5-CH=CH_2$	1642
$CH_3-(CH_2)_6-CH=CH_2$	1641
$ClCH_2-CH=CH_2$	1641
$BrCH_2-CH=CH_2$	1633
$HOCH_2-CH=CH_2$	1648

A frequency is termed *characteristic* (for a group of atoms) (*'group frequency'*) if it

— occurs in the spectra of all molecules that contain the structure element, and
— if it is not observed when this structure element is absent from the molecule (exceptions, p. 45).

The characteristic frequency regions are compiled in correlation tables (cf. Figs 3.1 and 3.2). These tables are an important tool for the interpretation of vibrational spectra.

As the next example illustrates, the use of correlation tables enables discrimination between classes of compounds by means of their vibrational spectra.

Example 3.2   Classification of a compound of known composition

A substance with the formula C_3H_6O could, in theory, belong to four different classes of compounds (Table 3.2). The distinction between an alcohol and an ether is easily made by comparing the characteristic frequency regions given in this table. The discrimination between an aldehyde and a ketone is not as simple.

---

† This is generally the case if the structure element is sufficiently different from the rest of the molecule (by 100% for one of the atomic masses or by 50% for the force constant).

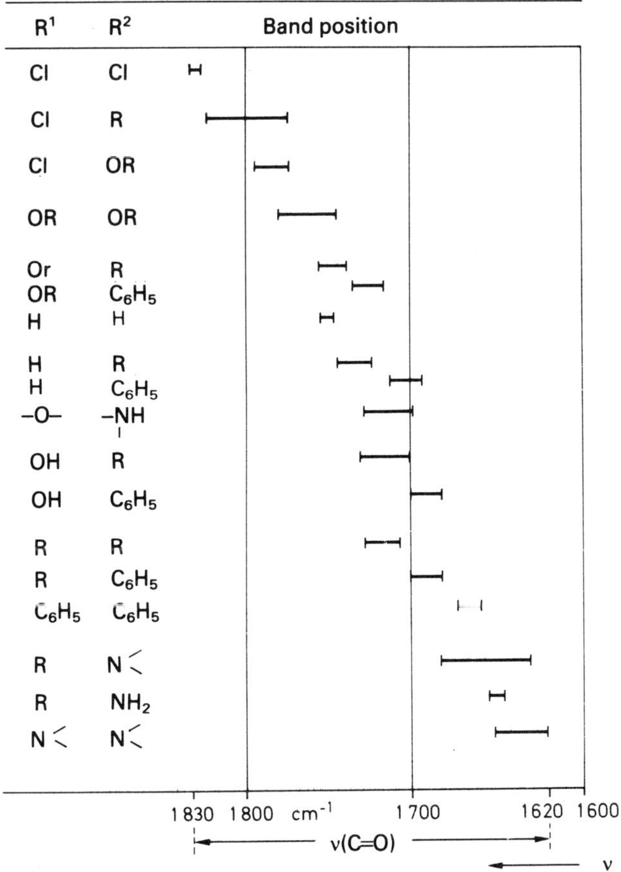

Fig. 3.2 — Exact band positions of several carbonyl compounds with the general formula

$$R^1 - \overset{\displaystyle O}{\overset{\|}{C}} - R^2$$

(R = aliphatic substituent).

In this case it is only possible to identify a carbonyl group (see Example 3.3 for a more detailed characterization).

The correlation tables can be used to formulate a dualistic question-and-answer scheme [11], which can readily be programmed.

The exact position of vibrational bands

It has been shown that the position of a band is roughly determined by the bond strength, the atomic masses and to a lesser extent by the neighbouring parts of the molecule. The exact position of these bands within a characteristic part of the

Table 3.2 — Classes of compounds and the corresponding wavenumber regions

Isomer	Class	Characteristic vibrations
$H_2C{=}CH{-}CH_2{-}OH$	Alcohol	$\nu(O{-}H)$ $\nu(C{=}C)$ $\nu(C{-}OH)$
$H_2C{=}CH{-}O{-}CH_3$	Ether	$\nu(C{=}C)$ $\nu(C{-}O{-}C)$
$H_3C{-}CH_2{-}\overset{\displaystyle O}{\underset{\displaystyle H}{C}}$	Aldehyde	$\nu(C{=}O)$
$H_3C{-}\overset{\displaystyle O}{\underset{\displaystyle }{C}}{-}CH_3$	Ketone	

$$4\,000 \quad cm^{-1} \qquad 3\,000 \qquad 2\,000 \qquad 1\,000 \qquad 0$$
$$\longleftarrow \nu$$

spectrum is, however, very much subject to influences from adjacent parts of the molecule.

> *Example 3.3   Determining the class of compound by detailed analysis of exact band position*

Carbonyl compounds usually exhibit a strong IR band between 1620 and 1830 cm$^-$1, which is characteristic of the C=O double bond. The exact position within this region depends on the inductive and mesomeric effects which the neighbouring groups exert on the C=O bond strength (Fig. 3.2). Knowledge of the exact band positions allows a more exact assignment of the compound class (e.g. [12]).

Less characteristic vibrations

In large molecules there has to be a differentiation between characteristic vibrations which are localized at one bond or functional group and those that involve larger parts of the molecule (section 2.1.1). Characteristic vibrations can be used to draw conclusions about the structure of different parts of the molecule. The frequencies involving larger parts of the molecule can shift considerably and are therefore less suitable for diagnostic purposes; but, as has been mentioned earlier, they are invaluable in compound identification (cf. finger-print spectra, Example 3.4).

    In summary then

— The interpretation of characteristic frequency regions allows the rapid identification of the functional groups of an unknown molecule and thus its classification. This is an important advantage of vibrational spectroscopy.

— Details, for example about the chemical environment of a functional group, can be derived from the exact band position.

— Finally, comparison with reference spectra, especially in the finger-print region, can lead to the identification of the compound.

Example 3.4   A simple scheme for the identification of an unknown compound

Step 1: The IR spectrum is recorded. The interpretation of the strongest absorption bands allows an initial classification.
    — characteristic regions:
        $v = 3000\text{–}2800 \text{ cm}^{-1}$; aliphatic CH_3/CH_2
        $v = 1700 \text{ cm}^{-1}$; carbonyl $C{=}O$
        $v = 1460\text{–}1380 \text{ cm}^{-1}$; CH_3/CH_2-deformation

Step 1: result: aliphatic carbonyl
    — take precise band positions into account:
        $v = 1714 \text{ cm}^{-1}$; ketone
    result: aliphatic ketone

Step 2: use other data, e.g. boiling point, to narrow down the choice of possible compounds

    b.p. $= 102°C$

    result: only the two isomers, methyl n-propyl ketone and and diethyl ketone, remain

Step 3: positive identification is achieved by a complete (!) comparison of all relevant spectra (**3**, Fig. 3.3). The most pronounced differences are observed in the finger-print region, e.g. between 1200 and 1300 cm^{-1} and around 1100 and 800 cm^{-1}.

Final result: the unknown substance is diethyl ketone.

Limitations of the method

The correlation of group frequencies cannot be used in some cases:

(1) Characteristic frequencies are present but indistinguishable:

— the usefulness of normally characteristic bands is diminished when several regions overlap.

Example 3.5   OH- and NH-stretching vibrations

The bands for $v\,(OH)$ and $v\,(NH)$ in the IR spectrum of 1-amino-3-hydroxypropane $H_2N\ (CH_2)_3\ OH$ (Fig. 3.4(a)) are not well separated. They occur in the same spectral region and have comparable intensities, so only one very broad band between 3600 and 3200 cm^{-1} is observed. Only the Raman spectrum (Fig. 3.4(b)) can identify the amino group, because $v_s\,(NH_2)$ ($v = 3299 \text{ cm}^{-1}$) and $v_{as}\,(NH_2)$ ($v = 3360 \text{ cm}^{-1}$) yield stronger (and sharper) Raman bands than the OH group.
    — The absence of a band in a characteristic region is not conclusive for the absence of the structure element. The corresponding vibration can be IR- and/or Raman-inactive.

Example 3.6   v(CC) in molecules of the type $X_2C{=}CX_2$

The characteristic band of the $C{-}C$ stretching vibration does not appear in the IR spectra of C_2X_4 compounds (e.g. C_2Cl_4). Group theoretical considerations lead

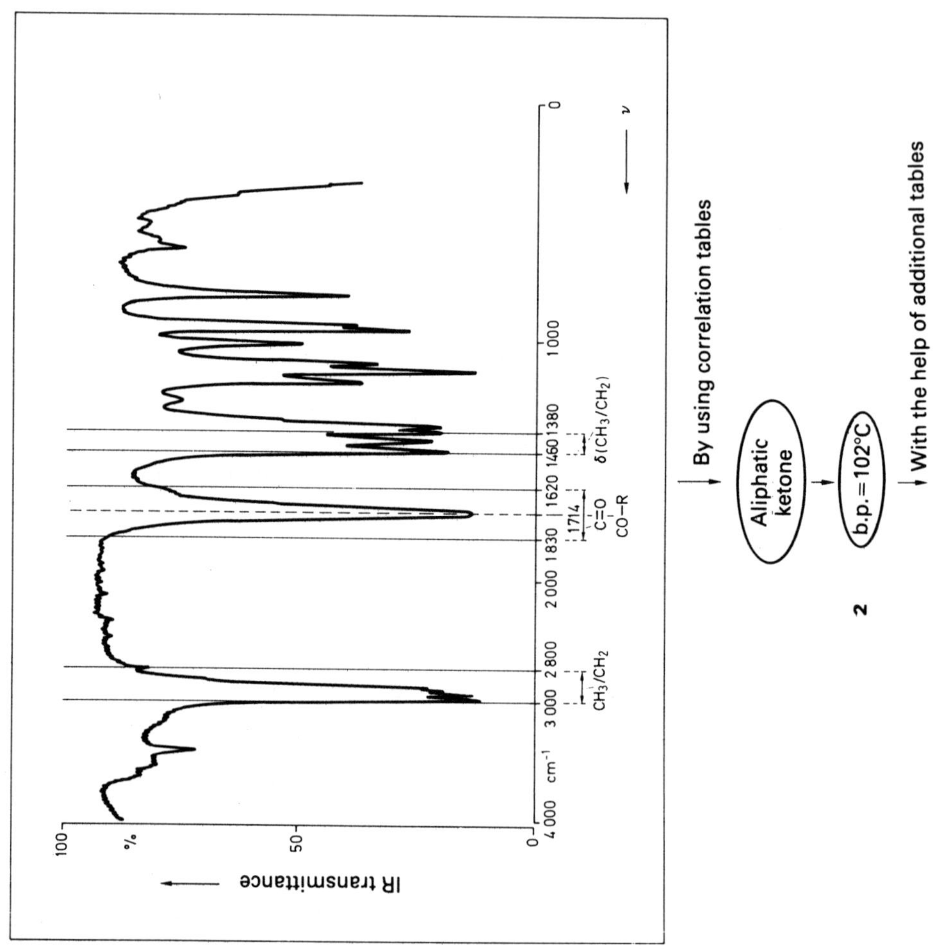

IR transmittance
%
100
50
0
4 000 cm⁻¹
3 000 2 800
2 000 1 830
1714
1 620
1460 1380
1 000
0
CH₃/CH₂
C=O
CO—R
δ(CH₃/CH₂)
ν
By using correlation tables
Aliphatic ketone
b.p. = 102°C
With the help of additional tables
2

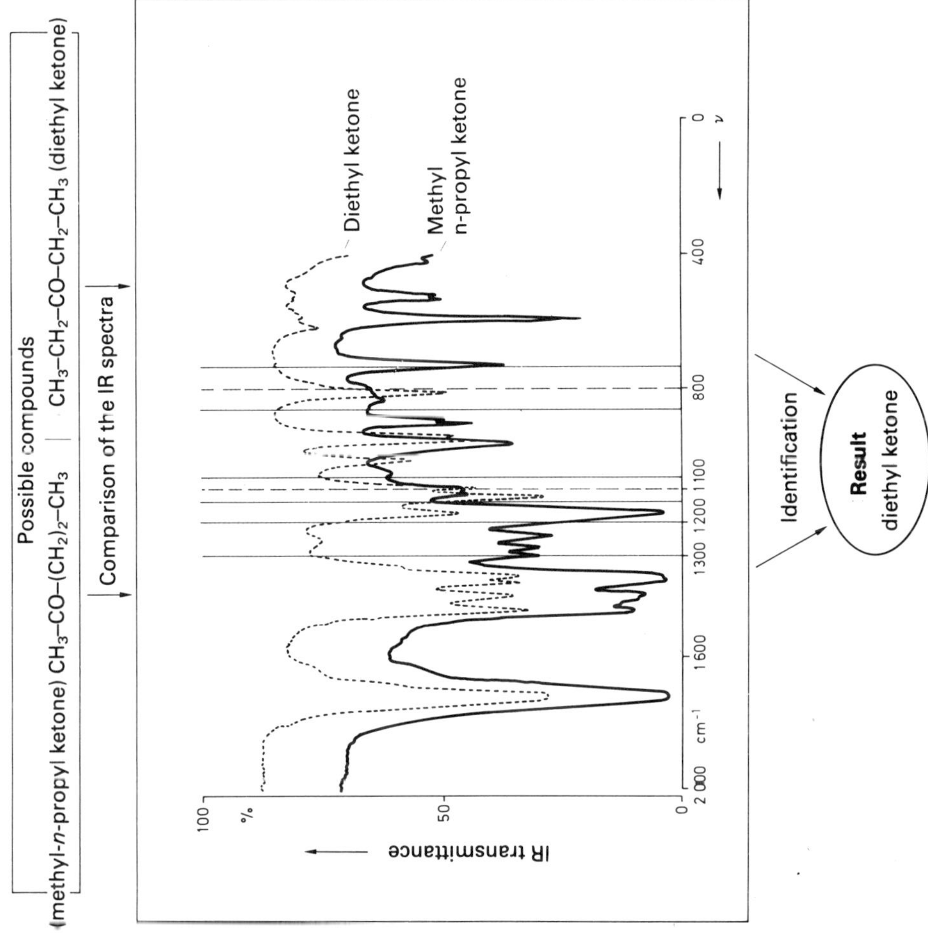

Fig. 3.3 — General scheme for the identification of an unknown compound: **1** — IR spectrum; interpretation of the strongest bands; **2** — additional physical data; **3** — comparison of all relevant spectra.

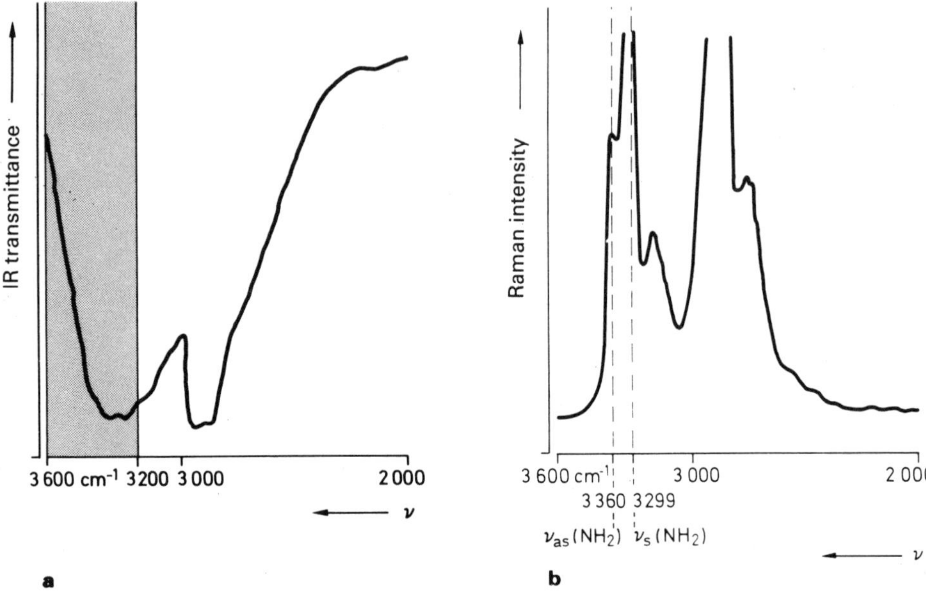

Fig. 3.4 — Sections of the (a) infrared and (b) Raman spectrum of 1-amino-3-hydroxypropane $(H_2N-(CH_2)_3-OH)$.

to the conclusion that this vibration is IR-inactive.

The expected band is, however, observed in the Raman spectrum of these compounds and allows their classification as olefins. The combination of Raman and IR data yields information about their symmetry (Example 1.3).

(2)   Other factors that can shift or eliminate group frequencies:

— Fermi resonance can cause a relatively small deviation from the expected position (Example 1.2), but the bands lose their identity. The interaction of a normal vibration with an overtone makes separate assignment impossible.
— Vibrational coupling. Well defined characteristic frequencies are observed for organic molecules, which can often be described as being composed of almost independent oscillators.

In many cases, inorganic molecules cannot be characterized in this way because of strong vibrational coupling, which occurs when the bond strengths and atomic masses vary only slightly throughout the molecule, and separation into independent structure units is impossible. Consequently, the concept of 'characteristic frequencies' should not be applied too rigorously for analytical purposes.

Example 3.7   Fundamental vibrations of the carbon dioxide molecule (CO_2)

The strong vibrational coupling of the two identical C=O units, through the common carbon atom, results in two stretching vibrations of different symmetries (Fig. 1.5(a)), which occur far outside the characteristic region of the isolated C=O-stretching frequencies.

isolated $C=O$ $\nu \approx 1830–1620\ cm^{-1}$
$O=C=O$ $\nu_{as} = 2349\ cm^{-1}$; $\nu_s \approx 1345\ cm^{-1}$,
(but split owing to Fermi resonance).

3.1.2   General appearance of the spectrum

The method of selective interpretation of a few (most intense) bands, which was described in section 3.1.1, leads in most cases to the classification and identification of an unknown compound. The general impression of a large region can also hint at the presence of structural features, such as functional groups, without requiring a detailed assignment. Table 3.3 shows the typical features of some organic compound classes.

3.2   INTERPRETATION OF BAND SHIFTS

The frequency ν'_{vib} of a vibration is determined by the bond strength and the atomic masses (section 3.1.1):

$$\nu \propto \nu'_{vib} \propto \sqrt{f(m_1 + m_2)/(m_1 m_2)} \qquad (3.2.1)$$

A shift of the wave number ν can therefore be caused by a change in:

— one of the atomic masses, with f = constant (isotopic molecules, cf. section 3.2.1)
— the bond strength, with m_i = constant (section 3.2.2)
— the atomic masses *and* bond strengths. This generally leads to new compounds and will not be discussed here.

In addition, with bond strength and masses constant, a band shift can be caused by changes in the coupling with other vibrations (section 3.2.3).

3.2.1   A change in one of the atomic masses, while the bond strength remains unchanged

An increase in one or both masses of the oscillator lowers the frequency of the stretching vibration; if the bond strength is not altered, the replacement of an atom A by its heavier isotope A′ does not change the bonding in the molecule; bond length and angles are the same, but the frequencies of all normal vibrations which involve A′ are reduced.

> *Example 3.8   Vibrational frequencies of hydrogen cyanide (HCN) and deuterium cyanide (DCN)*

All wave numbers of $D–C\equiv N$ are lower than those observed for $H–C\equiv N$. D and H are obviously involved in all normal vibrations (section 3.2.3). The shifts are of varying magnitude. The largest shift is observed for the C D stretching vibration. The isotopic shifts are an important aid for the assignment of frequencies. They are the basis for the isotope reduction method (Chapter 8) which is used to calculate force constants.

Table 3.3 — Characteristics of the IR spectra of some classes of compounds

Compound class	Feature	Example	IR spectrum
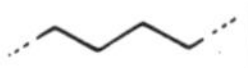 Aliphatic chains of at least four methylene groups	Intense bands between 3000 and 2800 cm^{-1} and bands around 1400 cm^{-1} plus a band at 720 cm^{-1}	Nujol (mixture of unbranched hydrocarbons)	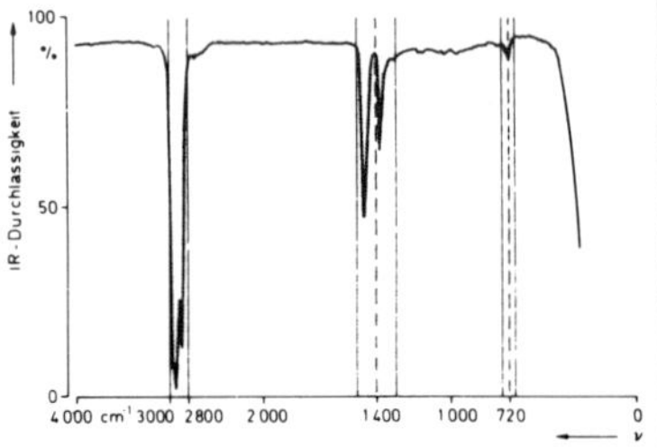 100 % IR - Durchlässigkeit 50 0; 4000 cm^{-1} 3000 2800 2000 1400 1000 720 0 ν
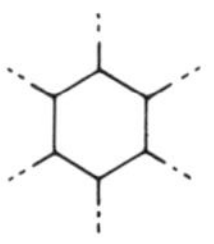 Aliphatic ring	Several evenly spaced bands of equal intensity between approx. 1300 and 800 cm^{-1}	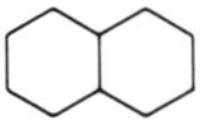 Decalin	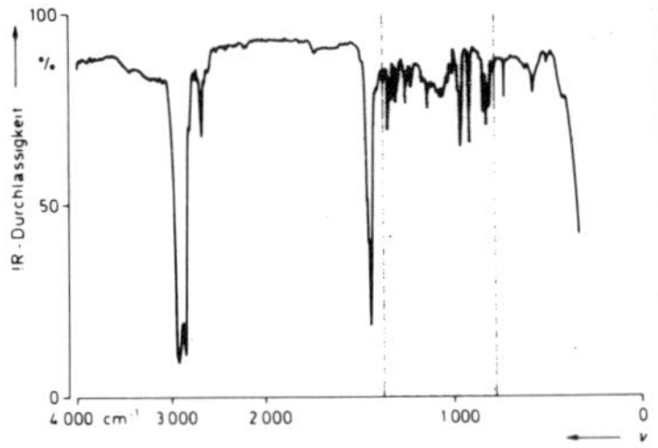 100 % IR - Durchlässigkeit 50 0; 4000 cm^{-1} 3000 2000 1000 0 ν
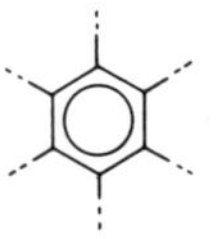 Aromatic compounds	Bands between 2000 and 1700 cm^{-1}; their number, position and shape are characteristic of the type of substitution (pronounced only at high sample concentrations)	 Toluene	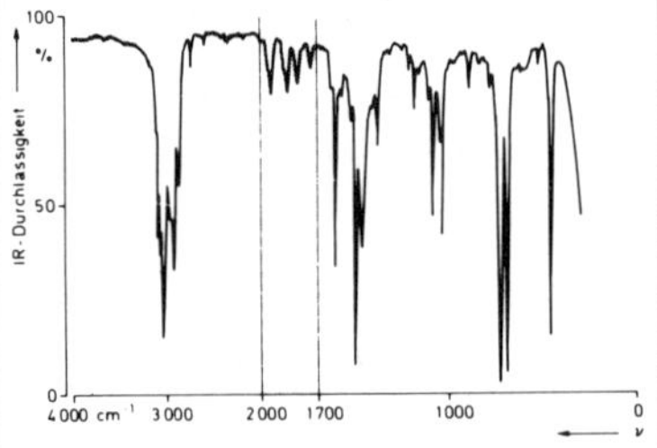 100 % IR - Durchlässigkeit 50 0; 4000 cm^{-1} 3000 2000 1700 1000 0 ν

3.2.2 Change of bond strength, with constant atomic masses

A stronger bond between two atoms increases the frequency of the corresponding stretching vibrations. This can be used to determine changes in the bonding caused by sorption, adduct or complex formation and to reach conclusions about co-ordination.

Table 3.3 — Continued

Compound class	Feature	Example	IR spectrum
Acid–amide peptides	Several bands of decreasing intensity between 1750 and 1100 cm^{-1} (saw-tooth type)	N-ethylacetamide	
Carboxylic acids	Broad band between 3000 and 2500 cm^{-1} and medium broad band between 960 and 875 cm^{-1}	Palmitic acid	
Amino acids	Broad, intense band between 3000 and 2300 cm^{-1} with some fine structure and a characteristically shaped band between 2200 and 2100 cm^{-1}	Glycine	

Sorption mechanisms

The separation of inorganic complexes is often carried out by using so-called complexing sorbents. Unlike classical ion exchangers, these sorb by co-ordination through a functional group (e.g. $-C=O$, $-C=S$, $-P=O$, $-CONH$, etc.) rather than by the exchange of a proton or another ion.

The changes in bonding during the sorption are readily monitored by vibrational spectroscopy. Valuable information is obtained by observing the direction of the band shift relative to the starting material.

Example 3.9   Sorption of tetrabromoauric acid $H[AuBr_4]$ on a xylanc/methoxy methylisocyanate polyurethane

Observation. Comparison of the spectra of pure (I) and loaded (II) polyurethane shows that $\nu(C=O)$ decreases, while $\nu(C-O)$ increases (Table 3.4).

Table 3.4 — Wave number shifts caused by sorption

Vibration	I	II	Δv
$v(C{=}O)$	1730	1711	-19
$v(C{-}O)$	1250	1270	$+20$

The C=O double bond is obviously weakened by the sorption.
Model. Theoretically, co-ordination by the urethane group is possible through three different atoms, or a combination of these.

In the first two cases one would expect a strengthening of the C=O bond through inductive effects and therefore an increase in $v(C{=}O)$. Co-ordination through the carbonyl oxygen atom reduces the C=O bond strength and thereby $v(C{=}O)$, which is in agreement with the observed shifts.

Note. An explanation of the observed changes in bonding requires a detailed theoretical study of all factors, e.g. changes of the molecular geometry, hybridization, etc. It should be stressed at this point that vibrational spectroscopy can provide information about the bonding and its changes, but it cannot give theoretical explanations.

Complex compounds
In analogy to the above it is possible to draw conclusions about the co-ordination in a complex. The analysis of vibrational spectra is especially advantageous for ligands which can be co-ordinated through different atoms. Again, the direction of the band shifts relative to the free ligand provides clues about the type of complex.

Example 3.10   Thiocyanato complexes

In principle, the thiocyanate ion can be co-ordinated through either the sulphur or the nitrogen atom:

(1) M—SCN

(2) M—NCS

The two types of complexes can be distinguished easily by their vibrational spectra. The type of co-ordination modifies the strength of the C−S and C−N bonds and thereby the corresponding stretching vibrations. This effect is illustrated by their mesomeric structures.

(1) co-ordination through the sulphur atom

$$M-\bar{\underline{S}}-C\equiv\bar{N} \qquad \longleftrightarrow \qquad M-\overset{+}{\underline{S}}=C=\bar{\underline{N}}^{\,-}$$

$$\textbf{1} \hspace{10em} \textbf{2}$$

(2) co-ordination through the nitrogen atom

$$M-\overset{+}{N}\equiv C-\bar{\underline{S}}\mathrm{I}^{\,-} \qquad \longleftrightarrow \qquad M-\bar{N}=C=\bar{S}$$

$$\textbf{3} \hspace{10em} \textbf{4}$$

Generally the resonance structure, in which the charges are separated, has little significance; **1** outweighs **2** in the case of co-ordination through the sulphur atom.

If co-ordination occurs through the nitrogen atom, then structure **4** is favoured. It can therefore be assumed that the C−S bond will be strengthened by the increase of its double-bond character and that the C−N bond will be weakened. Accordingly one expects:

$\nu(CS)$ (S co-ordination) $< \nu(CS)$ (N co-ordination) and

$\nu(CN)$ (S co-ordination) $> \nu(CN)$ (N co-ordination)

The type of thiocyanate complex can thus be easily derived from vibrational spectra (Table 3.5).

3.2.3 Band shifts caused by vibrational coupling

Frequency shifts cannot always be explained in terms of altered bond strength or atomic masses. Direct comparison of the vibrational frequencies of bonds or larger parts of a molecule is only possible if their normal vibrations are comparable. So they must have corresponding symmetries and an equal number of similar bonds and angles, which take part in the vibration.

Example 3.11   Hydrogen cyanide/deuterium cyanide ($H-C\equiv N$/$D-C\equiv N$)

The wavenumber of the C−N stretching vibration of DCN is $172\ \mathrm{cm}^{-1}$ lower than that of HCN (Table 3.6). Since the bond strengths of the isotopic molecules are identical and the masses involved (C,N) are unaffected by the substitution, there must be another factor causing the band shift.

The interaction between the C−D and the C−N stretching vibration is stronger, since $\nu(CD)$ is closer to the 'isolated' $\nu(CN)$ than $\nu(CH)$. The 'CN

Table 3.5 — Dependence of the vibrational wave numbers [cm^{-1}] on the type of thiocyanate complex [13]

Co-ordination through the	Characteristic regions		Examples		
	v(CN)	v(CS)	Complex	v(CN)	v(CS)
Sulphur atom	2200cm^{-1} 2000	900cm^{-1} 600	[Hg(SCN)$_4$]$^{2-}$	2120	719
			[Rh(SCN)$_4$]$^{2-}$	2106	705
			[Pd(SCN)$_4$]$^{2-}$	2108	703
			[Pt(SCN)$_4$]$^{2-}$	2114	700
			[Pt(SCN)$_4$]$^{2-}$	2120	694
Nitrogen atom	2200cm^{-1} 2000	900cm^{-1} 600	[V(NCS)$_4$]$^{2-}$	2077	830
			[Fe(NCS)$_4$]$^{2-}$	2050	828
			[Cr(NCS)$_4$]$^{2-}$	2092	825
			[Mo(NCS)$_4$]$^{2-}$	2071	820
			[Co(NCS)$_4$]$^{2-}$	2076	820
			[Zn(NCS)$_4$]$^{2-}$	2079	815
			[Ni(NCS)$_4$]$^{2-}$	2072	766

vibration' of DCN includes larger parts of the DC vibration than is the case for HCN; therefore v(CN) and v(CD) are 'coupled'. This results in a mutual repulsion of the two vibrational frequencies, which lowers the position of 'v(CN)' in DCN relative to HCN.

Table 3.6 — Potential energy distribution (PED) and wavenumbers [cm^{-1}] of the CN-stretching vibrations [14]

Molecule	'v(CN)'	Potential energy distribution	
		%CN	%CH w.r.t. CD
HCN	2097	95	5
DCN	1925	66	34

The bond strengths of the two molecules cannot be compared simply in terms of their vibrational spectra, as in section 3.2.2.

3.3 CRITERIA FOR THE INTERPRETATION OF VIBRATIONAL SPECTRA

The assignment of single bands to structure elements of a molecule (i.e. bonds, groups of atoms, or functional groups) is sufficient for the applications described in sections 3.1 and 3.2.

The more thorough analysis of vibrational spectra that follows assumes that each band in the spectrum can be assigned to a vibration of the molecule. The first step in the assignment is the decision about which bands belong to fundamental vibrations of the molecule. In most cases this starts with the strongest bands. If this selection is correct then all remaining frequencies can be derived from the combination of fundamental vibrations.

An accurate interpretation requires knowledge of the number, symmetry and activity of all normal vibrations, and, in addition, information about intensities and band shapes. Table 3.7 summarizes the parameters from which conclusions for the assignment can be drawn.

Table 3.7 — Requirements for and evidence derived from some criteria for the assignment

Material	Parameters	Information about	Task	Sections
One spectrum	Position, intensity, Contour, Degree of polarization (of a band)	Strength, polarity (of a bond or group); Symmetry of the vibration	Identification; Complete assignment	3.1.1, 3.3.1, 3.3.2, 3.3.3
	Overall impression	Structure elements	Identification	3.1.2
Several spectra	Spectra of similar molecules; Isotope shifts	Corresponding structure element	Complete assignment	3.2.1
Calculations	Force constants; Quantum theory	To check the assignment		

3.3.1 Intensities

The intensity of an IR absorption band is proportional to the change in the dipole moment μ during the vibration:

$$I(\text{IR}) \propto (\partial\mu/\partial Q)^2 \tag{3.3.1}$$

(Q = normal co-ordinate of the vibration). Similarly the intensity of a Raman band is proportional to the change in the polarizability α during the vibration:

$$I(\text{Ra}) \propto (\partial\alpha/\partial Q)^2 \tag{3.3.2}$$

These expressions are of little practical value since the magnitude of the dipole moment and polarizability, or their changes, are normally not known.

The following general rules can provide some criteria for estimates:

— Polar bonds yield intense IR and weak Raman bands.
— The opposite is true for weakly polar or non-polar bonds, for which weak IR and strong Raman bands are observed.
— The intensity also depends on the number of identical structure elements in a molecule.

Table 3.8 gives some examples for IR and Raman intensities of various structure elements. These should be taken only as points of reference and not as rules of universal validity.

It is obvious that the respective vibrations have to be IR/Raman-active (cf. Example 3.6).

3.3.2 Band contours

In Table 1.2 the excitation of a molecular vibration was represented schematically by the transition from the ground state ($v = 0$) to the first excited state ($v = 1$) ($\Delta v = +1$ for absorption). For a more precise description it is necessary to include the rotational states $J = 0,1,2,\ldots$, which are superimposed on every vibrational state (cf. Table 3.9(a)).

A molecule can exist in one of several rotational states in every vibrational state. The vibrational transitions therefore occur from the rotational states of the vibrational ground state to those of the first excited state. Simultaneous vibrational and rotational transitions are also possible, so that $\Delta v = +1$ can be combined with $\Delta J = 0$, $\Delta J = +1$ and $\Delta J = -1$.

For gases at low pressures, instead of a single band accordingly a series of sharp maxima is observed (rotation–vibration spectrum, Table 3.9(b)). At high pressures or low resolution only the broad envelope is observed (Table 3.9(c)). The purely vibrational transition ($\Delta v = +1$, $\Delta J = 0$) is labelled Q, the high-frequency part ($\Delta J = +1$) is termed R-branch and the other P-branch. Rotational structure is clearly observed only for gases. It is barely visible for liquids, owing to the hindered rotation, and completely absent from the vibrational spectra of solids.

Band contours and rotational structure can sometimes simplify the assignments of some bands. In linear molecules, for example, vibrations along the molecular axis are forbidden if $\Delta v = +1$, $\Delta J = 0$ and allowed if $\Delta v = +1$, $\Delta J = 1$. The vibrational spectra therefore show a minimum between the P- and R-branches. Vibrations perpendicular to the molecular axis exhibit bands with P-, Q- and R-structures.

Example 3.12   IR-band contours of acetylene (H−C≡C−H); (Table 3.9(d) and (e))

The anti-symmetric stretching vibration v_{as} (CH) occurs parallel to the molecular axis. The Q-branch is therefore absent from the corresponding IR band ($v = 3282$ cm^{-1}). The anti-symmetric deformation δ_{as} occurs perpendicular to the axis of the molecule. The Q-branch is clearly visible in the IR spectrum of this band.

Comment. This example is intended to illustrate the interpretation of the band shape only. The assignment of these two IR-active bands is already obvious from their positions.

Similar relationships between vibrations and band contours can be found for non-linear molecules (cf. Siebert [16]).

Table 3.8 — Rules of thumb for the approximation of intensities of IR and Raman bands for some structure elements [15]

System[†]	Structure element	Vibration	Intensity[‡] IR	Raman
X−Y	$-\overset{\textstyle\vert}{\underset{\textstyle\vert}{C}}-\overset{\textstyle\vert}{\underset{\textstyle\vert}{C}}-$	$v(C-C)$	w	s
	$-\overset{\textstyle\vert}{\underset{\textstyle\vert}{C}}-X^{*}$	$v(C-X)$	s	vs
	R−O−H	$v(O-H)$	s	w
X=Y	$\diagup C=C \diagdown$	$v(C=C)$	w	s
	$\diagup C-O$	$v(C=O)$	3	3
X≡Y	$-C\equiv C-$	$v(C\equiv C)$	w	s
Y−X−Y	O=C=O	$v_s(YXY)$	w−m	s
	$-\overset{\textstyle\vert}{\underset{\textstyle\vert}{C}}-O-\overset{\textstyle\vert}{\underset{\textstyle\vert}{C}}-$	$v_{as}(YXY)$	s	w
	$-N=C-N\diagup\diagdown$			

† X = heavy atom.
‡ w − weak, m − medium, s − strong, vs − very strong.

3.3.3 Depolarization ratio

The symmetry of a vibration influences certain parameters of the corresponding band, for example the band shape.

Bands belonging to totally symmetric vibrations† can be distinguished from others in the Raman spectra of gases, liquids and solutions by determining the depolarization ratio for every band.

† Vibrations are totally symmetric if all symmetry elements are preserved during their course. In group theory they form the first of the A-type irreducible representations (Chapter 5).

Table 3.9 — Schematic representation of rotation–vibration spectra and their basic transitions

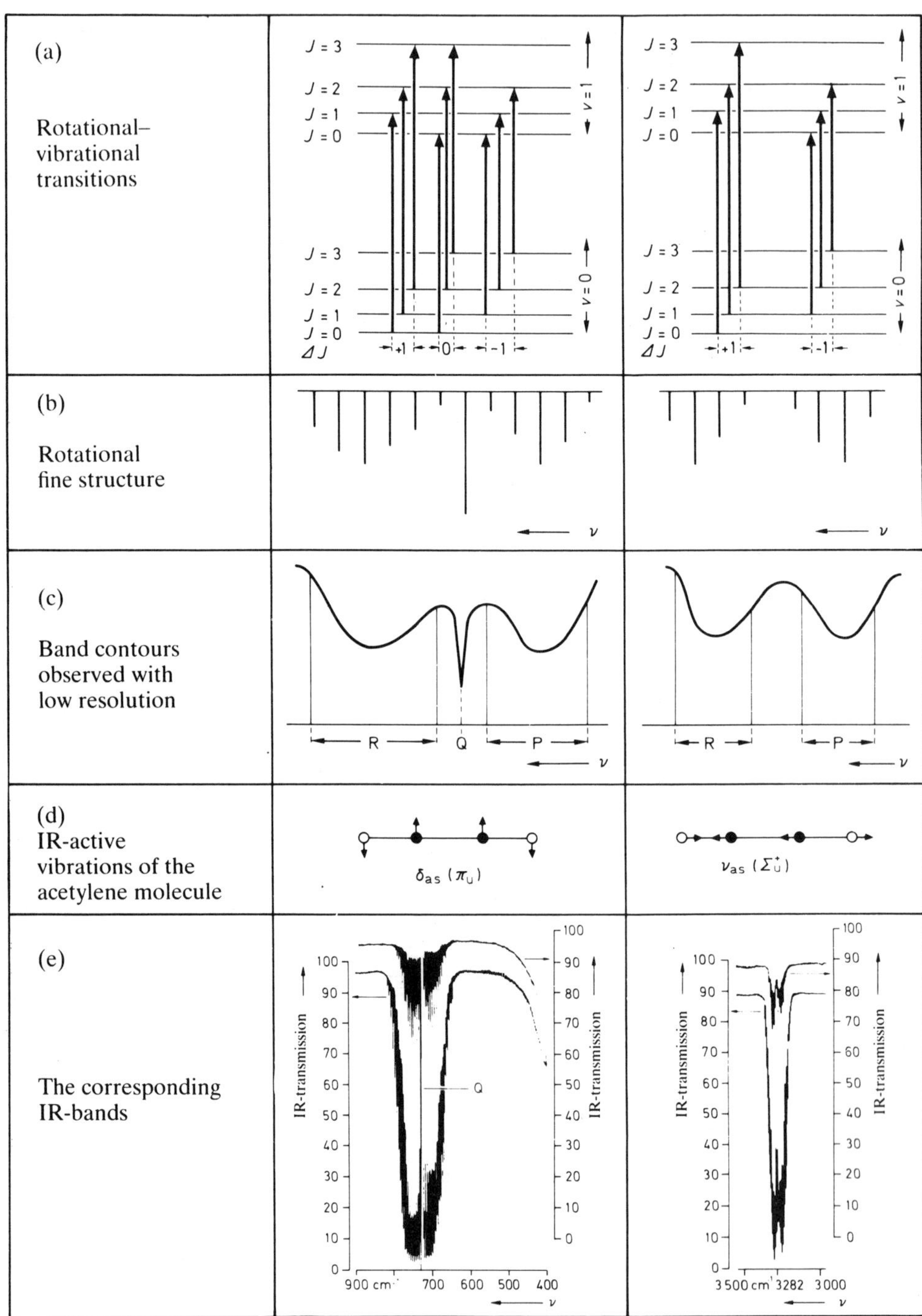

Definition

The depolarization ratio ρ is defined as the ratio of two intensities:

$$\rho = I_\perp / I_\parallel \qquad\qquad (3.3.3)$$

where $I_\parallel$, $I_\perp$ are the intensities of the scattered radiation with their plane of polarization parallel and perpendicular to that of the laser beam. To determine ρ, two Raman spectra (or the relevant regions) are recorded with planes of polarization of the incident light rotated through 90°, keeping all other parameters constant. The comparison of the intensities shows that $I_\perp$ and ρ are determined by the symmetry of the corresponding vibration (Fig. 3.5).

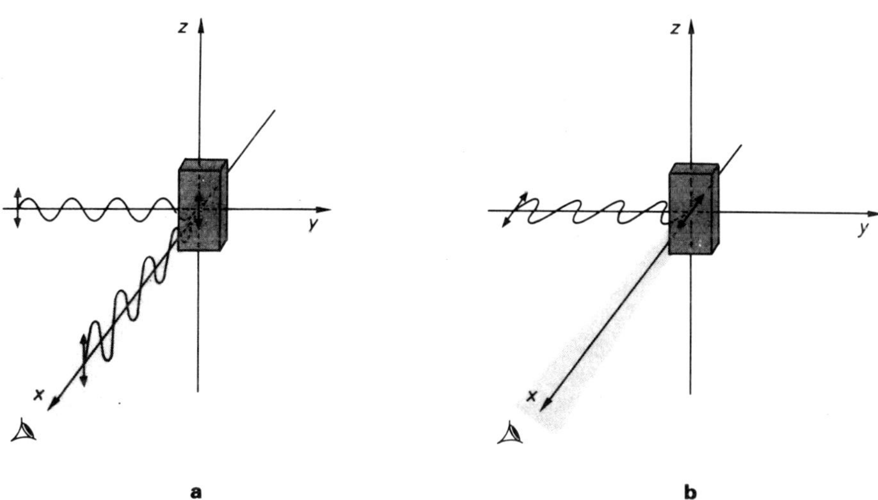

Fig. 3.5 — Determination of the depolarization ratio $\rho = I_\perp / I_\parallel$: (a) intensities observed parallel ($I_\parallel$) and (b) perpendicular ($I_\perp$) to the plane of polarization of the incident radiation.

Totally symmetric vibrations.   Electromagnetic radiation which is polarized in the yz-plane (Fig. 3.5(a)) induces a dipole oscillating in this plane. This molecular dipole can also emit radiation in the x-direction ($I_\parallel$). Turning the plane of polarization of the exciting radiation consequently turns the axis of the oscillating dipole. Since a dipole can emit light only parallel to its axis, no radiation will be detected in the x-direction in this case.

In theory $I_\perp$ and consequently ρ are zero. The original band will therefore have (almost) disappeared in the second measurement. Such a band is said to be 'polarized'.

Vibrations of low symmetry (anti-symmetric or degenerate).   For vibrations which destroy at least one symmetry element, $I_\perp = 3/4\, I_\parallel$ so that the theoretical value for ρ is 3/4.

Example 3.13   Depolarization ratios of the Raman bands of carbon tetrachloride
                (CCl_4)

Fig. 3.6 shows Raman spectra of carbon tetrachloride, which were recorded

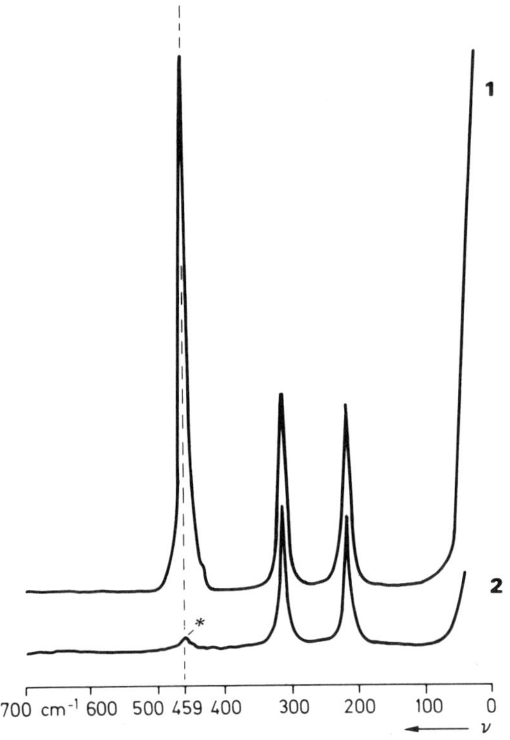

Fig. 3.6 — Depolarized (**1**) and polarized (**2**) Raman spectra of carbon tetrachloride. *polarized Raman band.

parallel ($I_\parallel$, trace **1**) and perpendicular ($I_\perp$, trace **2**) to the plane of polarization of the incident radiation. Two bands are only slightly weakened, whereas the band at $v = 459$ cm^{-1} has almost disappeared in the second measurement. Its depolarization ratio is almost zero, and it is therefore immediately assignable to a totally symmetric vibration of the molecule, v_s.

Comment. The following mathematical treatment shows, that for totally symmetric vibrations ρ is not exactly zero, but assumes values from 0 to 3/4. A differentiation of vibrations of low symmetry may then be more difficult.

The determination of the depolarization ratio allows the assignment of Raman bands to vibrations of different symmetry, but not the identification of structure elements.

Mathematical treatment
The depolarization ratio for linearly polarized light is

$$\rho = \frac{3(\gamma')^2}{45(\bar{\alpha}')^2 + 4(\gamma')^2} \tag{3.3.4}$$

where $\bar{\alpha}' = 1/3\ (\alpha'_{xx} + \alpha'_{yy} + \alpha'_{zz})$ contains only the derivatives of the diagonal elements of the tensor α (cf. section 1.3.3), whereas the anisotropy factor

$$(\gamma')^2 = 1/2\{(\alpha'_{xx} - \alpha'_{yy})^2 + (\alpha'_{yy} - \alpha'_{zz})^2 + (\alpha'_{zz} - \alpha'_{xx})^2 \\ + 6(\alpha'^2_{xy} + \alpha'^2_{yz} + \alpha'^2_{zx})\} \tag{3.3.5}$$

also includes derivatives of off-diagonal elements†. The polarizability ellipsoid (Fig. 1.6) can help to illustrate this relationship. Its changes will be determined by the symmetry of the vibration.

Totally symmetric vibrations. The polarizability ellipsoid changes its size. The derivatives of the diagonal elements of α are

$$\alpha'_{ii} \neq 0; \Lambda\ \bar{\alpha}' \neq 0 \Lambda\ 0 < \rho < 3/4$$

The Raman band is polarized.

Vibrations of lower symmetry. The polarizability ellipsoid does notchange its size; α_{xx}, α_{yy}, α_{zz} are constant:

$$\bar{\alpha}' = 0 \Lambda\ \rho = 3/4$$

The Raman band is depolarized.

3.4 SUMMARY

The basis for most applications and the principal task in the analysis of vibrational spectra is the assignment of Raman and IR bands. To be precise this means the assignment band $\rightarrow$ vibration. However, for some simplified interpretations the assignment band $\rightarrow$ structure element will be sufficient (cf. sections 3.1 and 3.2).

For a more precise assignment the intensity and shape of a band have to be considered in addition to its position. In special cases further material may be included, such as other spectra, isotope shifts, calculations, etc.

---

† The interested reader can find a derivation of Eqs. (3.3.4) and (3.3.5) in [17].

The fundamental question at the beginning of each assignment is whether the vibration considered is spectroscopically active. This decision can be made on the basis of symmetry and group theoretical considerations, which will form the basis for all advanced studies (Chapters 5 and 6).

4

Quantitative investigations

The quantitative analysis of systems containing more than one component is an important application area. An exact assignment band $\rightarrow$ vibration may not always be necessary. However, in contrast to the preceding chapters, some mathematical methods will be employed. The usefulness of the obtainable data is, as with other spectroscopical or physical methods, determined by the precision and reproducibility of the measurements, which are determined by:

— the preparation of samples:
  Gases are diluted with inert gases (nitrogen, noble gases). The measurements are always carried out at the same temperature and total pressure. Liquids should be measured in the same cuvette with a well defined path length. Solids are always studied in solution because measurements on discs are not reproducible because of variations in sample dispersion, pressing conditions, sample homogeneity and light scattering, etc. The availability of a solvent which does not interfere with the measurement is therefore the limiting factor for the applicability of the technique.
— the measurement:
  All parameters, e.g. scan speed, stress on the sample, etc., have to be kept constant.
— the analysis of the data:
  The absorbance $A(\lambda)$, which now replaces the older term extinction E, is the measure of the amount of electromagnetic radiation that is absorbed by a sample. Various graphical techniques for the quantitative analysis of spectra are in use. They can, however, yield different results for the same band. A meaningful comparison is possible only for values that have been obtained with the same method.

This chapter will give only a short description of the basic principles of the quantitative analysis of vibrational spectra.

In recent years many specialized methods for quantitative analysis have been developed, but only two general applications — the determination of unknown concentrations and the study of reaction kinetics — will be described in detail in this section.

4.1   ANALYSIS OF VIBRATIONAL SPECTRA

The IR spectrum is a graphical representation of the transparency of a sample for electromagnetic radiation as a function of the wave number v expressed in terms of the transmittance $\tau_i(\lambda)$ so that

$$\tau_i(\lambda) = f(v) \tag{4.1.1}$$

where $\tau_i(\lambda)$ is defined as the ratio of the intensities before (I_0) and after (I) the passage through the sample:

$$\tau_i(\lambda) = \frac{I}{I_0} \tag{4.1.2}$$

4.1.1   Band intensities and concentrations

The relationship between the transmittance $\tau_i(\lambda)$ and the molar concentration c (in mole/l) of the analyte in a sample of thickness d (in cm) is described by the Lambert–Beer law (Eq. (4.1.3)):

$$\tau_i(\lambda) = \frac{I}{I_0} = 10^{-xcd} \tag{4.1.3}$$

where x is a constant with units $1.\text{mole}^{-1}\text{cm}^{-1}$, which depends on the wavelength of the absorbed radiation but not on the concentration c. x is called the molar absorption coefficient or molar absorptivity (in the past it was called the extinction coefficient).

By rearranging Eq. (4.1.3) we obtain the absorbance $A(\lambda)$ which is also directly proportional to the concentration:

$$-\log \tau_i(\lambda) = \log \frac{I_0}{I} = xcd = A(\lambda) \tag{4.1.4}$$

$A(\lambda)$ is more useful for quantitative analysis than $\tau_i(\lambda)$ because it is linearly proportional to the concentration c, despite the requirement for the conversion of the spectral data. The absorbance $A(\lambda)$ is additive in the case of overlapping bands of several components with concentrations $c_1, c_2, \ldots, c_n$:

$$A(\lambda)_{\text{total}} = d\sum_{i=1}^{n} x_i(\lambda)c_i \qquad (4.1.5)$$

4.1.2 The determination of band intensities

For a completely accurate determination of band intensity, it is necessary to integrate over the entire area of the band. Since this is quite cumbersome it is common practice to measure only the absorbance at the band maximum to obtain $A(\lambda) = \log (I_0/I)_{\text{max}}$ (cf. Fig. 4.1). While I can be read directly from the trace, I_0 can be determined in

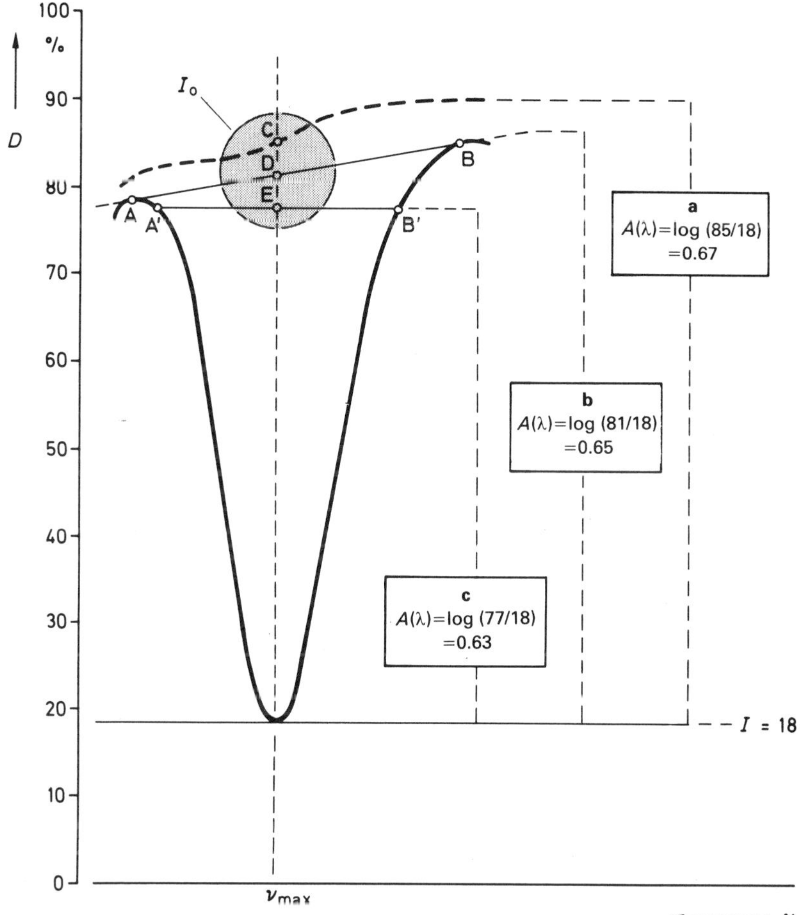

Fig. 4.1 — For the determination of the intensity I_0: **a** background absorption method (– – – solvent); **b** baseline method; **c** two constant reference points.

three different ways:

— The baseline is defined by the spectrum of the solvent which is recorded with the same spectrometer settings (method of background absorbance correction, **a** in Fig. 4.1).
— A straight line is drawn through two reproducible points of approximately the same height in the spectrum, e.g. absorption minima (baseline method point A and point B, **b** in Fig. 4.1).
— Two points on the spectral trace which are the same distance from v_{max}, regardless of the band shape, are joined by a straight line (selection of two constant reference points; $l(A'E) \simeq l(B'E)$, **c** in Fig. 4.1).

The value of I_0 corresponds to the intercept of one of the chosen reference lines with a vertical line through v_{max} (cf. Fig. 4.1): For the same absorption band it is thus possible to find different absorbances ($A(\lambda) = 0.67, 0.65$ and 0.63 respectively), since I_0 depends on the choice of reference line. Therefore only values that have been obtained by the same method are comparable. The analysis of the Raman spectrum is carried out correspondingly, but without conversion to a logarithmic scale and without the arbitrary choice of the method of quantification. For simple investigations, the height of the Raman peak is used as a measure of its intensity and the reference is given by the background of the spectrum which is, in the case of solutions, usually quite constant.

4.1.3   The determination of concentrations

To calculate an unknown concentration, as well as I and I_0, which can be derived from the spectrum, x and d have to be known. These do not have to be determined if several samples of known concentration are used to prepare a calibration graph by plotting the absorbances vs. the concentrations. The unknown concentration can then be found by interpolation.

Example 4.1   The determination of oil in refinery effluents [18]
Infrared spectroscopy is a quick and simple method for the determination of different types of oil and their concentrations in the waste water.

Sample preparation. The oil contaminants of 2500 ml of waste water are extracted with 50 ml of carbon tetrachloride (CCl_4). The spectra are recorded in 1-cm quartz cells with solvent compensation. For the quantitative analysis it is sufficient to measure the region of the C−H stretching vibrations, i.e. between 3300 and 2800 cm^{-1}.

Calibration graph. For the quantitative determination, three types of oil have to be distinguished since their molar absorptivities differ significantly. This gives rise to bands of varying intensities for the same overall oil concentration. For each type, varying concentrations of a reference compound were dissolved in water and extracted, to obtain the calibration graphs.

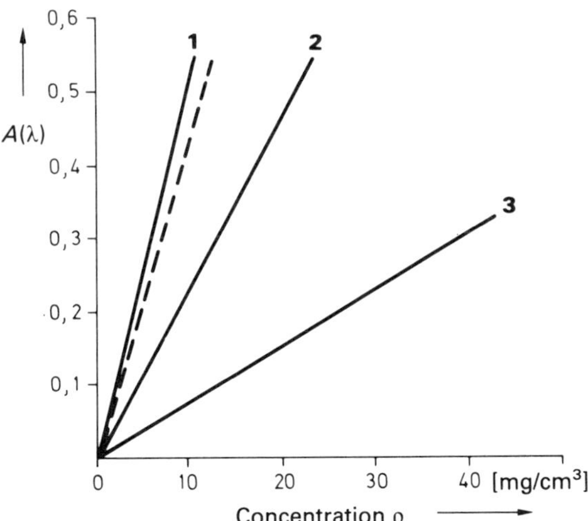

Fig. 4.2 — Calibration graphs for the determination of several types of oil in waste water from an oil refinery: **1** iso-octane; **2** isopropylbenzene; **3** benzene; — — — standard oil mixture (reproduced with permission of Perkin-Elmer & Co GmbH).

Type: pure aliphatic,   compound: iso-octane **1**
        pure aromatic,   compound: benzene **2**
        aliphatic-aromatic,   compound: isopropylbenzene **3**

A standard oil mixture (37.5% iso-octane, 37.5% hexadecane, 25% benzene) was treated in the same way.

$$(H_3C)_2CH-(CH_2)_4-CH_3 \qquad\qquad \qquad\qquad CH(CH_3)_2$$

1   **2**   **3**

Fig. 4.2 shows the calibration lines for the three compounds and the standard oil mixture. To use these calibrations in an analysis it is necessary to know only the type of the compound in the sample, not the compound itself. In most cases this can be determined from the position and structure of the CH-bands.

Table 4.1 — The rearrangement of the penta-amine cobalt nitrito complex into the penta-amine cobalt nitro complex. Changes in the infrared spectrum: **1** after 4 days at ambient temperature; **2** after 6 days at ambient and 1 day at 60°C

	Penta-amine complex nitrito complex	Penta-amine cobalt nitro complex
Formula	$[Co(NH_3)_5ONO]^{2+}$	$[Co(NH_3)_5NO_2]^{2+}$
Structure	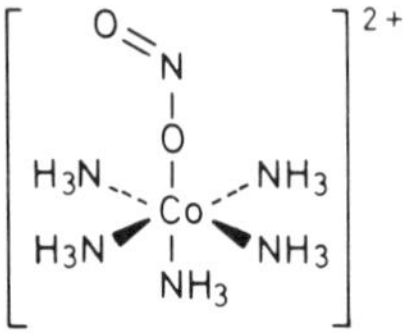	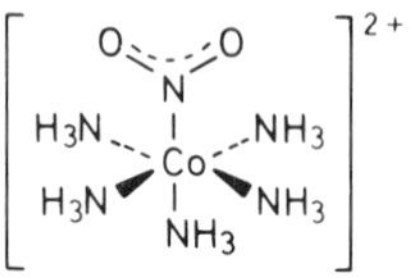
Infrared spectrum of the pure compound	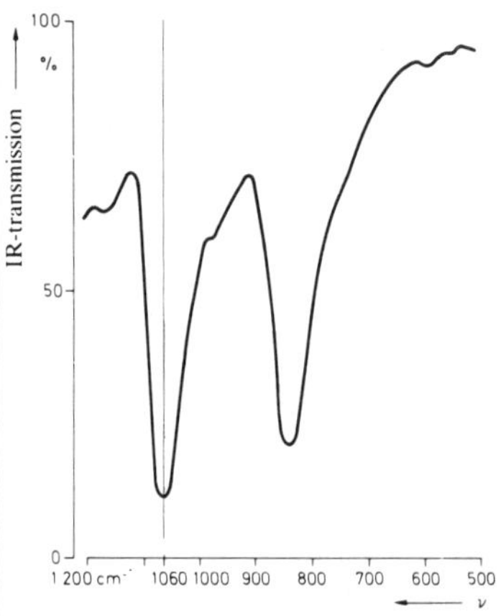	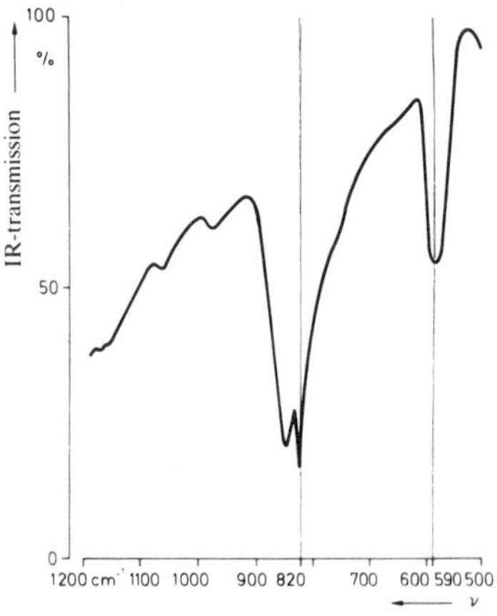
Rearrangement	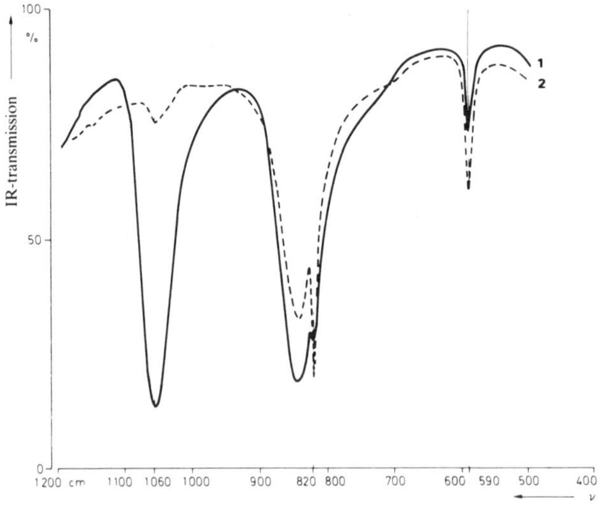	

4.2 KINETIC INVESTIGATIONS

In studies of the kinetics of a reaction, the concentrations of one or more of the reactants or products are determined at known intervals. This is possible with vibrational spectroscopy since the intensities of infrared or Raman bands are proportional to the concentrations.

Several methods can be used for this purpose:

— Samples (gases, liquids, solids) are taken continuously, prepared and analysed.
— Solutions or gases are pumped through an infrared or Raman cuvette which is connected to the reaction apparatus. The flow is then stopped for the time of the measurement (the quantitative analysis of these spectra is an extension of Example 2.3 in which the changes occurring in a spectrum were used to control a reaction).
— In the case of rearrangement reactions a spectrum is recorded at the start of the reaction. The measurement is then repeated at regular intervals (cf. Example 4.2). Since it is always the same sample it is also possible to study pelleted samples.

In the first two cases a new sample has to be taken for every measurement. Since the amounts of sample can vary, a standard has to be included.

Example 4.2   Nitrito → nitro rearrangement in a complex [19]
The complex described by the formula $[Co(NH_3)_5NO_2]^{2+}$ can exist in two isomeric forms (cf. Table 4.1):

— in the nitrito isomer, co-ordination occurs through an oxygen atom of the nitro group;
— in the nitro isomer, co-ordination takes place through the nitrogen atom.

The symmetry and bonding of the NO_2-group in the two isomers are different, so they can be distinguished by their vibrational spectra (see Table 4.1 for the spectra of the pure isomers). The bands at $1060\,cm^{-1}$ (nitrito isomer), 820 and $590\,cm^{-1}$ (nitro form) can be used for identification.

Quantitative investigation of the rearrangement. A KBr pellet is made immediately after the preparation of the penta-amine cobalt nitrito-complex $[Co(NH_3)_5ONO]^{2+}$. Infrared spectra are recorded at one- to two-day intervals, to follow the rearrangement into the nitro form: The intensity of the band at $1060\,cm^{-1}$ decreases rapidly and the bands at 820 and $590\,cm^{-1}$ increase steadily. The absorbance $A(\lambda)$ of the band at $1060\,cm^{-1}$ is determined as a function of time according to $A(1060)_t = \log\,[I_0(1060)/I(1060)]$. Finally the disc is heated for several hours at 100°C (to reach equilibrium) to obtain $A(\lambda)_\infty$.

Determination of the order of the reaction. There are three ways of plotting the absorbances as a function of time:

$$A(\lambda) - A(\lambda)_\infty = f(t)$$

$$\frac{1}{A(\lambda) - A(\lambda)_\infty} = f(t)$$

$$\log[A(\lambda) - A(\lambda)_\infty] = f(t)$$

but only the last of these gave a straight line. The reaction is therefore first order (cf. standard texts on physical chemistry). The rate constant k can be derived from the slope of the line ($k = 2.33 \times 10^{-6}\,\mathrm{sec}^{-1}$).

Part III
Interpretation of vibrational spectra with use of group theory

The topics of this section are the determination of molecular symmetry and applications based on this knowledge, such as the discrimination between isomers and modifications. Special emphasis is placed on the correspondence between the symmetry of a molecule, the symmetry of its vibrations and their activities. These considerations require not only a knowledge of symmetry, but also the application of group theory to molecular vibrations. They finally give rise to the formulation of 'selection rules'.

5

Group theoretical basis

This chapter describes the fundamental concepts of group theoretical methods and their formalisms. For many practical purposes, these can be applied without a detailed understanding of the mathematical background.

5.1 BASIC CONCEPTS AND DEFINITIONS

The symmetry of a molecule is described by symmetry operations and the corresponding symmetry elements.

5.1.1 Symmetry operations and symmetry elements

A symmetry operation (rotation, reflection, etc.) is an operation which moves an object into a new orientation equivalent to its original one. A symmetry element is a point, line, or plane with respect to which a symmetry operation is performed. Symmetry elements and symmetry operations are often denoted by the same symbol. Group theoretical analysis is only possible if a clear distinction is made between them, since several operations can belong to one element.

Two fundamentally different kinds of symmetry operation and element exist:

— closed symmetry operations which can be applied to a 'free' molecule and do not shift its centre of gravity, and

— open symmetry elements which include the translation of a structure unit and thereby describe its repetition throughout a crystal.

To begin with, we look only at free molecules. At most, five types of symmetry elements and corresponding operations are required for a complete description (Table 5.1):

— The neutral element E (sometimes also I for identity) is required only to satisfy certain mathematical requirements. The identity operation E does nothing to the molecule. This element and operation exist for every molecule.

— The axis of rotation C_n (symmetry element) exists, if the rotation (operation) of the molecule about this axis moves it into a new orientation which is equivalent to the preceding one. n denotes the order of the rotation and tells us what fraction of

Table 5.1 — The five closed symmetry operations and elements, and their application to the octahedral system AB_6 (the B-atoms have been numbered for clarity, although in reality they are, of course, indistinguishable)

Symmetry element	Symbol	Symmetry operation	Example
Neutral element	E	Identity operation	
Axis of rotation	C_n	Rotation by $\psi = (360/n)°$	C_4
Mirror plane	σ	Reflection through a plane	σ_h σ_v σ_d
Centre of inversion (center of symmetry)	i	Inversion through a point	
Improper axis of rotation	S_n	Rotation by $\psi = (360/n)°$ followed by reflection through a mirror plane perpendicular to this axis	S_4

The application of symmetry operations

a complete rotation we are to perform ($n = 360/\psi$). By convention the sense of the rotation is anti-clockwise. The superscript m tells us the number of rotations to be carried out in succession, e.g. C_3^2 is equivalent to a rotation by $2[360/3]° = 240°$. If several different axes of rotation occur in a molecule, then the axis with the highest order is called the principal axis.

Special cases:

C_1 denotes a rotation by $\psi = (360/1)° = 360°$ which is identical to the operation E,

C_∞ denotes a rotation by an infinitely small angle about the molecular axis. This can occur in linear molecules.

— A mirror plane divides the molecule into two identical parts. The reflection through this plane interchanges the two halves. There is a distinction between horizontal mirror planes h, which are perpendicular to the principal axis, and vertical mirror planes v, which are collinear with it. The dihedral mirror plane d is also collinear with the principal axis, but bisects the angle between two C_2-axes.

— A centre of inversion i. Inversion involves passing each atom through the point i at the centre of the molecule and placing it on an equivalent position on the opposite side. i does not have to coincide with an atom, e.g. i in H_2 lies in the middle between the two atoms.

— An improper axis of rotation S_n consists of an axis of rotation combined with a mirror plane perpendicular to it.

Special cases:

S_1 is identical to σ (rotation by 360° followed by a reflection)

S_2 is identical to i (rotation by 180° followed by a reflection)

5.1.2   Systematic description of symmetry

The SCHOENFLIES notation for symmetry elements was introduced in section 5.1.1. This notation is preferred in spectroscopy and many other areas of chemistry, whereas crystallographic applications use almost exclusively the HERMANN–MAUGUIN system, which is recommended by the International Union of Crystallographers. The important difference, apart from the fact that other symbols are used, is the replacement of S_n by:

— a rotation inversion axis, which involves a rotation by $\psi = (360/n)°$ followed by an inversion.

The symbol for this symmetry element and operation is $\bar{n}$, n being the order of the axis.

Special cases:

$\bar{1}$ is identical to i,

$\bar{2}$ is equivalent to a rotation by 180° followed by inversion, and thus identical to a reflection.

In Table 5.2 the two notations are listed for comparison.

Table 5.2 — Comparison of the Schoenflies and Hermann–Mauguin notations for symmetry operations

Symmetry operation	Schoenflies	Hermann–Mauguin
Rotation by $\psi = (360/n)°$	C_n	n
Reflection	σ	m
Inversion	i	$\bar{1}$
Rotation reflection	S_n	—
Rotation inversion	—	$\bar{n}$

5.1.3 Point groups†

The successive application of symmetry operations in molecules with more than one symmetry element will also result in an equivalent orientation of the molecule. The combination of symmetry operations gives the same result as the application of a single operation (cf. Example 5.1). It is generally true that:

— the product (successive application) of two symmetry operations is equivalent to another symmetry operation.

Example 5.1   The effect of symmetry operations on the octahedra AB_6

The reflection through the horizontal mirror plane h followed by inversion i will lead to a configuration which can also be obtained by a 180° rotation about the vertical axis:

$$i\sigma_n = C_2$$

(The operations in the product are always carried out from right to left!)

The symmetry operations which apply to any particular molecule collectively define the symmetry group of the molecule. In our case it is called a point group, since none of its operations will shift the centre of gravity of the molecule. The inclusion of open symmetry operations for solids will give rise to space groups.

---

† Point groups can be specified by using either Schoenflies or Hermann–Mauguin notation (cf. sections 5.1.1 and 5.1.2), but since spectroscopic studies generally use the Schoenflies notation, it will be used in this book.

These are not valid for the isolated molecule but for the whole crystal (cf. section 5.8).

5.2 GROUPS

Axioms of group theory. A set G consisting of the elements $A, B, C, \ldots$ forms a group, if:
(a) the product of any two elements is also an element of the set $C = A \cdot B$;
(b) the product is associative: $(A \cdot B) \cdot C = A \cdot (B \cdot C)$;
(c) there exists an element E (neutral element) in G which commutes with all other elements and leaves them unchanged: $A \cdot E = E \cdot A = A$ (commute means that the order of multiplication is not important);
(d) there exists in the group an inverse element for each element of the set G. Any element and its inverse must commute, so that $A \cdot X = X \cdot A = E$ then $X = A^{-1}$ and $A \cdot A^{-1} = A^{-1} \cdot A = E$.

Example 5.2 Integers
Does the set of positive and negative integers including zero form a group with respect to addition (1) and multiplication (2)?
(1) Addition
    (a) The number $m_3 = m_1 + m_2$ is an element of the group for all m_1 *and* m_2.
    (b) The addition is associative: $(m_1 + m_2) + m_3 = m_1 + (m_2 + m_3)$.
    (c) 0 is the neutral element of the group: $m_1 + 0 = 0 + m_1$.
    (d) The inverse element for m_1 is $-m_1$: $m_1 + (-m_1) = 0$.
    All four axioms are met and this set is therefore a group with respect to addition.
(2) Multiplication
    (a) $m_3 = m_1 \cdot m_2$ is an element of the set for all m_1 and m_2.
    (b) The multiplication is associative: $(m_1 \cdot m_2) \cdot m_3 = m_1 \cdot (m_2 \cdot m_3)$.
    (c) 1 is the neutral element: $m_1 \cdot 1 = 1 \cdot m_1 = m_1$.
    (d) The inverse element to m_1 *is* m_1^{-1}; but the set only contains integers. So all inverse elements except, $m_1^{-1} = 1$ are not elements of the set.
    The set therefore does not form a group with respect to multiplication.

These axioms do not apply only to groups of numbers. The complete set of symmetry operations which can be applied to a molecule meets these requirements and thereby defines its point group.

Example 5.3 Symmetry operations
Do the symmetry operations of the molecule thiazyltrifluoride (NSF_3) (Table 5.3) form a group?
(a) The product of two symmetry operations is identical to another operation of the set, e.g.

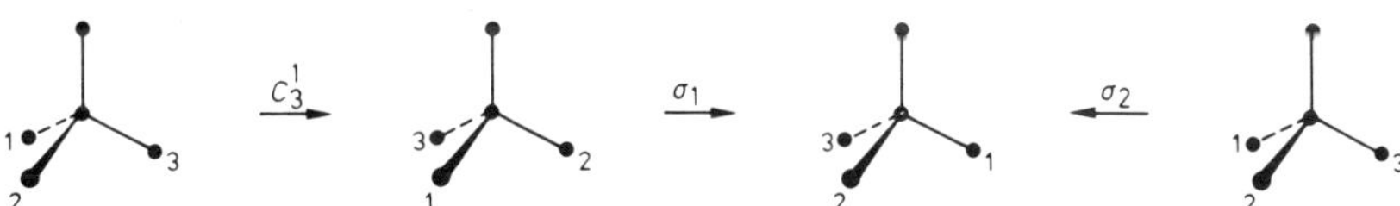

Table 5.3 — Symmetry elements and operations of NSF_3

Symmetry elements	Symmetry operations
E	E
C_3	C_3^1, C_3^2
σ_1	σ_1
σ_2	σ_2
σ_3	σ_3

Notation for the multiplication:

$$\sigma_1 \cdot C_3^1 = \sigma_2$$

The operations are applied in sequence from right to left. Table 5.4 contains the products of first (1st row) and second symmetry operation (1st column).

Table 5.4 — Multiplication table

	E	C_3^1	C_3^2	σ_1	σ_2	σ_3
E	E	C_3^1	C_3^2	σ_1	σ_2	σ_3
C_3^1	C_3^1	C_3^2	E	σ_3	σ_1	σ_2
C_3^2	C_3^2	E	C_3^1	σ_2	σ_3	σ_1
σ_1	σ_1	σ_2	σ_3	E	C_3^1	C_3^2
σ_2	σ_2	σ_3	σ_1	C_3^2	E	C_3^1
σ_3	σ_3	σ_1	σ_2	C_3^1	C_3^2	E

(b) The multiplication is associative, e.g.

$$(\sigma_1 \cdot \sigma_2) \cdot \sigma_3 = C_3^1 \cdot \sigma_3 = \sigma_2$$
$$\sigma_1 \cdot (\sigma_2 \cdot \sigma_3) = \sigma_1 \cdot C_3^1 = \sigma_2$$

(c) The identity operation is the neutral operation of the group, e.g.

$$E \cdot C_3^1 = C_3^1 \cdot E = C_3^1$$

(d) For each element of the group there is an inverse element so that their product is the neutral element E, e.g.

$$C_3^1 \cdot C_3^2 = C_3^2 \cdot C_3^1 = E$$

$$\sigma_1 \cdot \sigma_1 = \sigma_1^2 = E$$

The set of symmetry operations satisfies all group axioms and is called a symmetry point group.

It must be stressed that the elements of the point group are the symmetry operations and not the symmetry elements of the molecule.

5.3 THE DETERMINATION OF THE POINT GROUP

The first step in any of the following interpretations is the determination of the point group. This can be done by following a simple question-and-answer scheme if the symmetry elements† are known. Zeldin developed a scheme, which was extended by Salthouse and Ware to include tetrahedral, octahedral and icosahedral groups. The correct point group is found by following a path according to the presence (———) or absence (– – – cf. Fig. 5.1) of symmetry elements.

An example of such a scheme is the algorithm given in Fig. 5.1(a): The path begins with the separation into linear and non-linear molecules; after several branching points, the symbol of the appropriate point group is reached.

Example 5.4   Determination of the point group
The application of the algorithm is demonstrated in Fig. 5.1(b)–(e) for the cases of thiazyltrifluoride, **2** (NSF_3; 5.1(b)), carbon tetrachloride, **3** (CCl_4; 5.1(c)), benzene, **4** (C_6H_6; 5.1(d)) and hydrogen cyanide, **5** (HCN; 5.1(e)).

The effects that symmetry operations have on the positions of atoms are numerically classified and tabulated for most point groups. These tables are called 'character tables', because they contain 'irreducible characters' which are derived from group theory. These are the key to the discussions that follow, since they allow the determination of the following parameters:

— the symmetry types of the vibrations; section 5.4
— the IR-active modes of vibration; section 5.5.1
— the Raman-active modes of vibration; section 5.5.2
— the number of normal modes of vibration per symmetry species; section 5.5.3
— the activity of combinations of vibrations; section 5.6.1
— the activity of overtones; section 5.6.2

Together they determine the number of bands in the spectra, the coincidences of IR

---

† Even though the point groups are defined by the symmetry operations, it is often simpler to use symmetry elements for this purpose. This is possible because of the definite relation that exists between them.

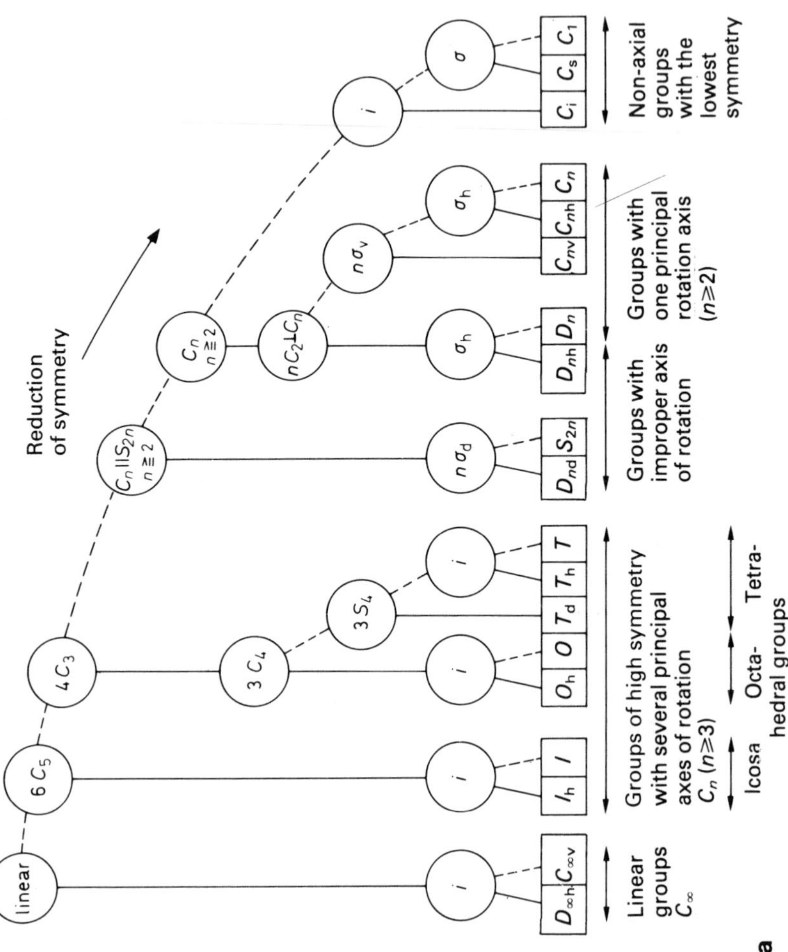

Fig. 5.1 — Determination of the point group: (a) the algorithm and its application to (b) NSF_3, (c) CCl_4, (d) C_6H_6, (e) HCN
C_n denotes the axis of rotation of the highest order; ———— denotes the presence and - - - the absence of symmetry elements.

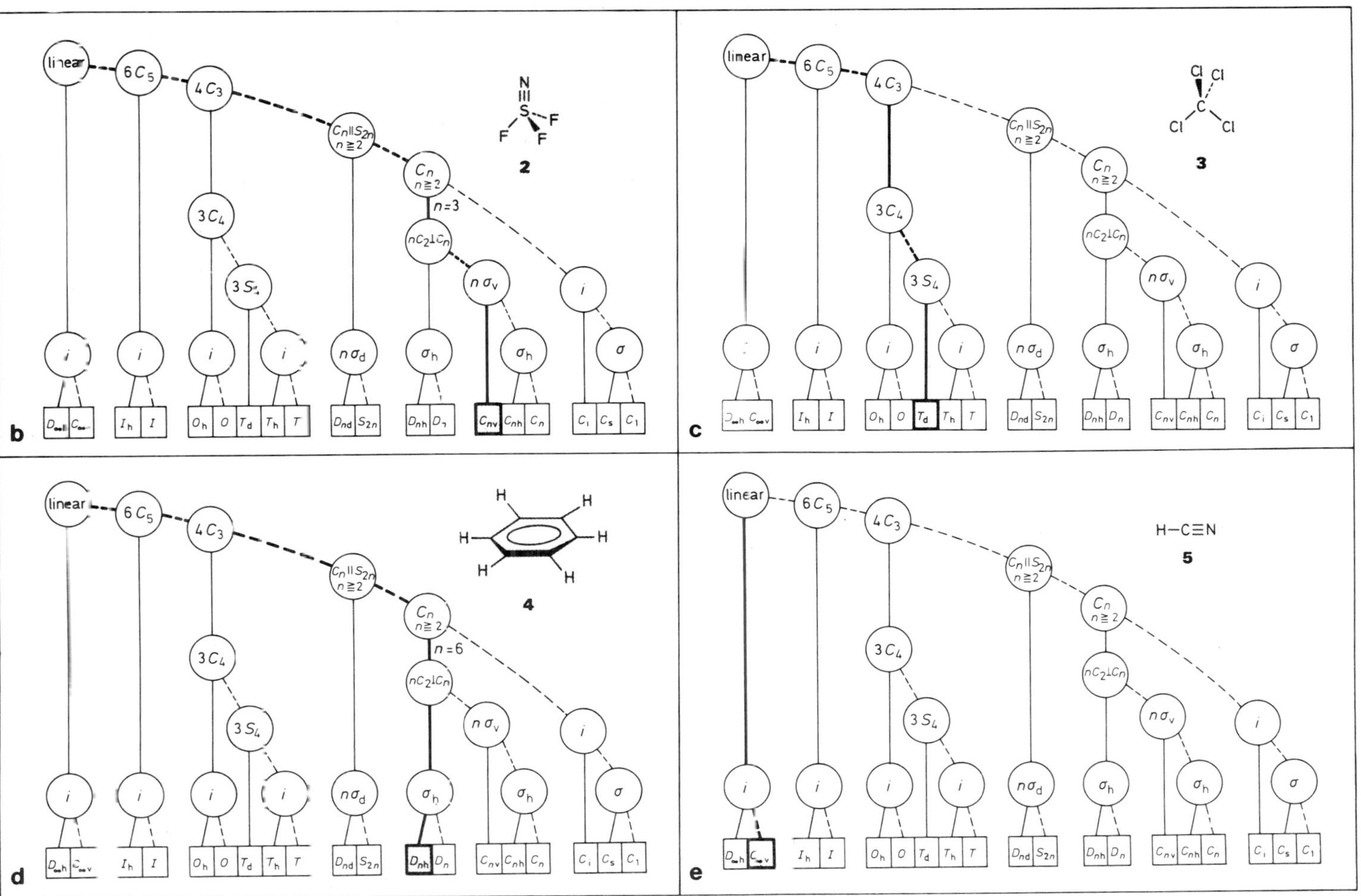

and Raman bands. They make the connection between the symmetry of a molecule and the spectral activity of its vibrations.

5.4   A GUIDE TO CHARACTER TABLES

The character table of the C_{3v} point group is shown in Fig. 5.2. Two points must be made:

(1) The symmetry operations of a system can be combined into *classes*: all operations in a class are equivalent.
(2) The normal vibrations of a molecule can be assigned to *symmetry species* (irreducible representations): all vibrations of a species are transformed in the same way by each symmetry operation. The character table (Fig. 5.2) contains information about:

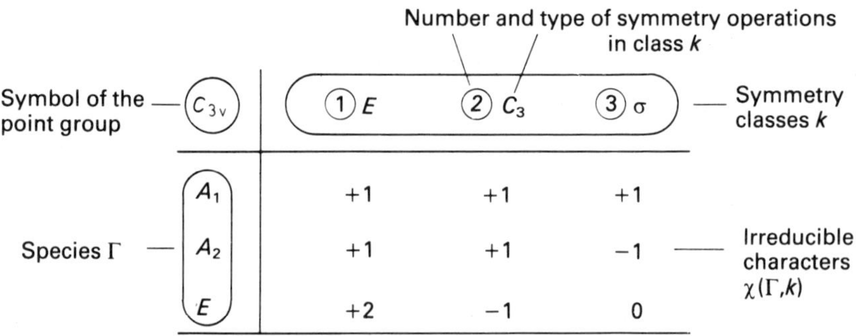

Fig. 5.2 — The structure of a character table, e.g. for the point group C_{3v}.

— Column 1: classification of the symmetry species to which the normal vibrations can belong. Each species differs from all others with respect to at least one symmetry operation. The notation is given in Table 5.5.
— Row 1: number $g(k)$ and type of the symmetry operations R of the point group, combined into classes k.
— between Row 1 and Column 1: the array of irreducible characters $\chi(\Gamma,k)$:

$\chi(\Gamma,k) = +1$:   a vibration belonging to the species is symmetrical with respect to the symmetry operations of the class k, i.e. the corresponding symmetry element remains intact throughout the entire vibration.

$\chi(\Gamma,k) = -1$:   the symmetry element is lost during the vibration. A symmetry operation of the class k would reverse a vibration belonging to the symmetry species Γ.

Table 5.5 — The meaning of the symbols used in character tables

Symbol		Meaning	
	Non-linear molecules		
Capital letters	A	Symmetrical	with respect to rotation about the principal axis
	B	Anti-symmetrical	
	E	Doubly	
	F	Triply	
	G	Quadruply	degenerate
	H	Quintuply	
Subscripts	1	Symmetrical	with respect to a second symmetry operation (e.g. C_2, σ_v)
	2	Anti-symmetrical	
	g	Symmetrical	with respect to inversion i
	u	Anti-symmetrical	
Superscripts	$'$	Symmetrical	with respect to σ_h
	$''$	Anti-symmetrical	
	Linear molecules		
	σ^+, Σ^+	Symmetrical	with respect to σ_v
	σ^-, Σ^-	Anti-symmetrical	
	π	Doubly degenerate	

$\chi(\Gamma,E) = +2, +3, \ldots$: the degree of degeneracy of the vibration.

$\chi(\Gamma,k) = 0$: follows from the requirement for orthogonality.†

---

† The g symmetry operations of a molecule, by analogy with the vectors of a g-dimensional space. From the definition of the g independent ('orthogonal') and scaled vectors

$$\mathbf{a \cdot a = g} \quad \text{and} \quad \mathbf{a \cdot b = 0}$$

one can derive the following conditions:

$$\sum_i \chi(\Gamma_i,k) \cdot \chi(\Gamma_i,k) = g \qquad \text{for each class}$$

$$\sum_i \chi(\Gamma_i,k_1) \cdot \chi(\Gamma_i,k_2) = 0 \qquad \text{for two classes } k_1 \text{ and } k_2$$

$$\sum_j \psi(\Gamma,k_j) \cdot \chi(\Gamma,k_j) \cdot g(k_j) = g \qquad \text{for each species}$$

$$\sum_j \chi(\Gamma_1,k_j) \cdot \chi(\Gamma_2,k_j) \cdot g(k_j) = 0 \qquad \text{for two species } \Gamma_1 \text{ and } \Gamma_2$$

5.5   SELECTION RULES I: NORMAL MODES OF VIBRATION

The $3N$–6 or $3N$–5 possible fundamental vibrations of an N-atomic molecule are distributed, in dependence on their symmetry, over the species Γ_i (section 5.4). A general reduction formula (Eq. (5.5.1)) can be used to determine:

— which species are infrared and/or Raman-active, and
— how many vibrations belong to each species.

$$Z(\Gamma) = [1/\sum_k g(k)] \cdot \sum_k g(k) \cdot \chi_{red}(k) \cdot \chi(\Gamma,k)] \tag{5.5.1}$$

The meaning of the reducible character $\chi_{red}(k)$ varies according to the parameter (Table 5.6) that has to be determined. The other parameters remain constant and can be taken directly from the character tables. $g(k)$ denotes the number of symmetry operations in class k, and $\chi(\Gamma,k)$ is the irreducible character of the symmetry class k of the species. The summation is extended over all symmetry classes k of the molecule.

Table 5.6 — The application of the reduction formula (Eq. 5.5.1))

Parameter	Meaning of $\chi_{red}(k)$	Relevant equations	Significance of $Z(\Gamma)$ $Z(\Gamma)=0$　　$Z(\Gamma)\neq0$
IR activity	Character of the dipole moment $\chi_\mu(k)$	Section 5.5.1	The species is IR-inactive　　IR-active
Raman activity	Character of the polarizability $\chi_\alpha(k)$	Section 5.5.2	The species is Raman-inactive　　Raman-active
Number of normal vibrations per species	$\theta(k)$	Section 5.5.3	There are no　　　　n normal vibrations in this symmetry species
Activity of combinations of Γ_1 and Γ_2	$[\chi(\Gamma_1,k)\cdot\chi(\Gamma_1,k)]$	Section 5.6.1	The combination is active if one of the species with $Z(\Gamma)\neq0$ is active
Activity of an overtone of the species Γ	$\chi_v(\Gamma,k)$	Section 5.6.2	The vth overtone is active if one of the species with $Z(\Gamma)\neq0$ is active

5.5.1   IR activity

To determine the IR activity of a vibration, $\chi_\mu(k)$, the character of the dipole moment will take the place of $\chi_{red}(k)$ in Eq. (5.5.1).

$$\chi(k) = 2\cos\psi + 1, \text{ for proper rotations } (E,C_n)$$

$$\chi(k) = 2\cos\psi - 1, \text{ for improper rotations } (\sigma,S_n,i).$$

ψ denotes the angle of rotation:

$$\psi = 0 \text{ for the operations } E \text{ and } \sigma,$$

$$\psi = (360/n)^\circ \text{ for } C_n \text{ and } S_n \text{ and}$$

$$\psi = 180^\circ \text{ for } i = S_2.$$

If $Z(\Gamma)$ is not zero, then the vibrations belonging to the symmetry species Γ will be IR-active. The magnitude of $Z(\Gamma)$ is not important.

5.5.2   Raman activity

To determine the Raman activity of a vibration, $\chi_\alpha(k)$, the character of the polarizability will take the place of $\chi_{red}(k)$ in Eq. (5.5.1).

$$\chi_\alpha(k) = 2 + 2 \cos \psi + 2 \cos (2\psi) \text{ for proper rotations and}$$

$$\chi_\alpha(k) = 2 - 2 \cos \psi + 2 \cos (2\psi) \text{ for improper rotations.}$$

5.5.3   The number of normal vibrations for any symmetry species

In this case $\theta(k)$ is substituted for $\chi_{red}(k)$. $\theta(k)$ is independent of m_k, the number of atoms which are not shifted by the symmetry operations of the class k.

$$\theta(k) = (m_k - 2) \, \chi_\mu(k) \text{ for proper rotations and}$$

$$\theta(k) = m_k \chi_\mu(k) \text{ for improper rotations.}$$

$Z(\Gamma)$ gives the number of normal vibrations per species Γ. The number of IR and Raman bands belonging to normal vibrations are obtained by combining the results of sections 5.5.1 and 5.5.2.

Example 5.5   Selection rules for thiazyltrifluoride (NSF_3)

How many bands, belonging to the nine normal vibrations of NSF_3, are to be expected in the IR and Raman spectra? (For the symmetry of NSF_3 see Example 5.3.)

(1) Determination of the point group (cf. Fig. 5.1(b)): C_{3v}

(2) Look up the corresponding character table (see also Fig. 5.2)

Table 5.7 — Character table of the point group C_{3v} (irreducible characters)

C	E	$2C_3$	$3\sigma_v$
A_1	$+1$	$+1$	$+1$
A_2	$+1$	$+1$	-1
E	$+2$	-1	0

(3) *Calculation of the parameters* $\chi_\mu(k)$, $\chi_\alpha(k)$ *and* $\theta(k)$:

Table 5.8 — Reducible characters

	E	$2C_3$	$3\sigma_v$
ψ	0	$+120$	0
$2\cos(\psi)$	0	-1	$+2$
$\chi_\mu(k)$	$+3$	0	$+1$
$2\cos(2\psi)$	$+2$	-1	$+2$
$\chi_\alpha(k)$	$+6$	0	$+2$
m_k	5	2	3
$\theta(k)$	$+9$	0	$+3$
$g(k)$	$+1$	$+2$	$+3$

(4) *Selection rules*

— IR activity:
$$Z(\Gamma) = \left[1\Big/\sum_k g(k)\right]\sum_k [g(k)\cdot\chi_\mu(k)\cdot\chi(\Gamma,k)]$$

— Raman activity:
$$Z(\Gamma) = \left[1\Big/\sum_k g(k)\right]\sum_k [g(k)\cdot\chi_\alpha(k)\cdot\chi x(\Gamma,k)]$$

— Number of bands per symmetry species
$$Z(\Gamma) = \left[1\Big/\sum_k g(k)\right]\sum_k [g(k)\cdot\theta(k)\cdot\chi(\Gamma,k)]$$

Determination of the IR activity:

$$Z(A_1) = \frac{1}{6}(1\cdot3\cdot1 + 2\cdot0\cdot1 + 3\cdot1\cdot1) = 1$$

$$Z(A_2) = \frac{1}{6}(1\cdot3\cdot1 + 2\cdot0\cdot1 + 3\cdot1\cdot(-1)) = 0$$

$$Z(E) = \frac{1}{6}(1\cdot3\cdot2 + 2\cdot0\cdot(-1) + 3\cdot1\cdot0) = 1$$

Determination of the Raman activity:

$$Z(A_1) = \frac{1}{6}(1\cdot6\cdot1 + 2\cdot0\cdot1 + 3\cdot2\cdot1) = 2$$

$$Z(A_2) = \frac{1}{6}(1\cdot6\cdot1 + 2\cdot0\cdot1 + 3\cdot2\cdot(-1)) = 0$$

$$Z(E) = \frac{1}{6}(1\cdot6\cdot2 + 2\cdot0\cdot(-1) + 3\cdot2\cdot0) = 2$$

Determination of the number of bands per species:

$$Z(A_1) = \frac{1}{6}(1\cdot9\cdot1 + 2\cdot0\cdot1 + 3\cdot3\cdot1) = 3$$

$$Z(A_2) = \frac{1}{6}(1\cdot9\cdot1 + 2\cdot0\cdot1 + 3\cdot3\cdot(-1)) = 0$$

$$Z(E) = \frac{1}{6}(1\cdot9\cdot2 + 2\cdot0\cdot(-1) + 3\cdot3\cdot0) = 3$$

Table 5.9 summarizes the results of the calculations.

Table 5.9 — Selection rules for NSF_3

| C_{3v} | Activity† | | | |
	IR	Raman	n_a	n_b
A_1	+	+	3	3
A_2	−	−	0	0
E	+	+	3	6
Sum			6	9

† + active, − inactive, n_a number of bands, n_b number of vibrations; $n_b = 3N - 6$ or $3N - 5$; if degeneracy occurs: $n_a < n_b$.

The vibrations of the species A_1 and E are IR- and Raman-active. Six fundamental bands are expected in both spectra (corresponding to $3 \times 5 - 6 = 9$ normal vibrations, since those belonging to the species E are doubly degenerate).

5.6 SELECTION RULES II: COMBINATIONS AND OVERTONES

5.6.1 Combinations

To determine whether a combination of two vibrations is allowed, i.e. active, it is necessary to multiply the irreducible characters of the two species Γ_1 and Γ_2 for each symmetry class k and then substitute the products $[\chi(\Gamma_1,k)\chi(\Gamma_2,k)]$ for $\chi_{red}(k)$ in the reduction formula (Eq. (5.5.1)).

Example 5.6   Combination of two vibrations of the octahedron AB_6

Are the combinations of vibrations belonging to the symmetry species F_{2g} and F_{1u} IR- or Raman-active (Table 5.10)?

Table 5.10 — Distribution of the 15 normal vibrations of the octahedron AB_6

	E	$8C_3$	$6C_2$	$6C_4$	$3C_2$	i	$6S_4$	$8S_6$	$3\sigma_h$	$6\sigma_d$	IR	Ra	n_a	n_b
A_{1g}	+1	+1	+1	+1	+1	+1	+1	+1	+1	+1	−	+	1	1
A_{2g}	+1	+1	−1	−1	+1	+1	−1	+1	+1	−1	−	−	0	0
E_g	+2	−1	0	0	+2	+2	0	−1	+2	0	−	+	1	2
F_{1g}	+3	0	−1	+1	−1	+3	+1	0	−1	−1	−	−	0	0
F_{2g}	+3	0	+1	−1	−1	+3	−1	0	−1	+1	−	+	1	3
A_{1u}	+1	+1	+1	+1	+1	−1	−1	−1	−1	−1	−	−	0	0
A_{2u}	+1	+1	−1	−1	+1	−1	+1	−1	−1	+1	−	−	0	0
E_u	+2	−1	0	0	+2	−2	0	+1	−2	0	−	−	0	0
F_{1u}	+3	0	−1	+1	−1	−3	−1	0	+1	+1	+	−	2	6
F_{2u}	+3	0	+1	−1	−1	−3	+1	0	+1	−1	−	−	1	3

Sum		15

Step 1: multiplication of the characters (Table 5.11)

Table 5.11 — Multiplication

	E	$8C_3$	$6C_2$	$6C_4$	$3C_2$	i	$6S_4$	$8S_6$	$3\sigma_h$	$6\sigma_d$
$\chi(F_{2g},k)$	+3	0	+1	−1	−1	+3	−1	0	−1	+1
$\chi(F_{2u},k)$	+3	0	+1	−1	−1	−3	+1	0	+1	−1
Product	+9	0	+1	+1	+1	−9	−1	0	−1	−1

Product $= [\chi(F_{2g},k)\cdot\chi(F_{2u},k)]$

Step 2: substitution into the reduction formula (Eq. 5.5.1)

$$Z(A_{1g}) = \frac{1}{48}(1{\cdot}9{\cdot}1 + 8{\cdot}0{\cdot}1 + 6{\cdot}1{\cdot}1 + 3{\cdot}1{\cdot}1 + 1{\cdot}(-9){\cdot}1 + 6{\cdot}(-1){\cdot}1 + 8{\cdot}0{\cdot}1 +$$
$$+ 3{\cdot}(-1){\cdot}1 + 6{\cdot}(-1){\cdot}1) = 0, \text{ etc.}$$

Result:

	A_{1g}	A_{2g}	E_g	F_{1g}	F_{2g}	A_{1u}	A_{2u}	E_u	F_{1u}	F_{2u}
Z	0	0	0	0	0	1	0	1	1	1

The combination of F_{2g} and F_{2u} is allowed, if $Z(\Gamma) \neq 0$ for at least one of the components determined by reduction. A_{1u}, E_u, F_{1u} and F_{2u} are Raman-inactive, but F_{2u} is IR-active; the combinations are therefore observed only in the IR spectrum.

5.6.2 Overtones

A similar method is used to determine whether the vth overtone of a normal vibration, belonging to species Γ, is active. In this case $\chi_v(\Gamma,k)$ is substituted for $\chi_{red}(k)$ in the reduction formula (Eq. (5.5.1)). But $\chi_v(\Gamma,k)$ is determined by multiplication as above only if the normal vibration is not degenerate.

$$\chi_v(\Gamma,k) = [\chi(\Gamma,k)]^v$$

More complicated formulae have to be used for degenerate vibrations. These are collected in Table 5.12.

5.7   A SUMMARY OF THE SELECTION RULES

The formulae used for the determination of selection rules are summarized in Table 5.13. This survey again illustrates the point that they are on the one hand based on symmetry parameters and on the other hand on formalisms derived from group theory.

5.8   EXTENSION OF SYMMETRY CONCEPTS TO THE SOLID STATE

The basics of symmetry and group theoretical methods described in sections 5.1–5.7 hold for isolated, 'free' molecules. As soon as they form part of a crystal lattice, the influences of neighbouring particles in the symmetry analysis must also be included.

In addition to the *closed symmetry operations*, which act on one molecule without

Table 5.12 — Reducible characters $\chi_\upsilon(\Gamma,k)$ of the υth overtone of the degenerate modes of vibration $\Gamma = E, F, G, H$ for the symmetry operations k

Species Γ $\chi_\upsilon(\Gamma,k)$

E

$$\frac{1}{2}\left[\chi^{\upsilon-1}(E,k)\cdot\chi(E,k) + \chi(E,k^\upsilon)\right]$$

F

$$\frac{1}{3}\left[2\chi^{\upsilon-1}(F,k)\cdot\chi(F,k) + \frac{1}{2}\chi^{\upsilon-2}(F,k)\cdot\{\chi(F,k)^2 - \chi(F,k^2)\} + \chi(F,k^\upsilon)\right]$$

G

$$\frac{1}{4}\left[3\chi^{\upsilon-1}(G,k)\cdot\chi(G,k) - \chi^{\upsilon-2}(G,k)\cdot\{\chi(G,k)^2 - \chi(G,k^2)\}\right.$$

$$\frac{1}{6}\chi^{\upsilon-3}(G,k)\cdot\{\chi(G,k)^3 - 3\chi(G,k)\cdot\chi(G,k^2) + 2\chi(G,k^3)\}$$

$$\left.\chi(G,k^\upsilon)\right]$$

H

$$\frac{1}{5}\left[4\chi^{\upsilon-1}(H,k)\cdot\chi(H,k) - \frac{3}{2}\chi^{\upsilon-2}(H,k)\cdot\{\chi(H,k)^2 - \chi(H,k^2)\}\right.$$

$$\frac{1}{3}\chi^{\upsilon-3}(H,k)\cdot\{\chi(H,k)^3 + 3\chi(H,k)\cdot\chi(H,k^2) + 2\chi(H,k^3)\}$$

$$\frac{1}{24}\chi^{\upsilon-4}(H,k)\cdot\{\chi(H,k)^4 - 6\chi(H,k)^2\cdot\chi(H,k^2) +$$

$$\left. + 8\chi(H,k)\cdot\chi(H,k^3) + 3\chi(H,k^2)^2 - 6\chi(H,k^4)\} + \chi(H,k^\upsilon)\right]$$

$$\chi_2(\Gamma,k) = \frac{1}{2}\left[\chi^2(\Gamma,k) + \chi(\Gamma,k^2)\right] \text{ for } \Gamma = E, F, G, H$$

$$\chi_0(\Gamma,k) = 1 \quad \text{for } \Gamma = E, F, G, H$$

$$\chi_\upsilon(\Gamma,k) = 0 \quad \text{for } \Gamma = E, F, G, H$$
for all negative values of υ

shifting its centre of gravity, there are two operations which act on several molecules and include translation. They are called '*open' symmetry operations* since the starting point of the operation is not reached again. Formally they are composed of a closed operation and a translation:

— *Combination of reflection and translation: glide reflection*   A glide plane is present when every atom of a molecule A is transformed into an equivalent atom on molecule A′ by means of a reflection followed by a translation parallel to the

Table 5.13 — The application of the selection rules

Section rules for		Activity IR RA	Normal vibrations	Combinations Overtones	
Prerequisites		Point group Geometry of the dipole moment | polarizability	Point group and molecular structure	Point group but not molecular structure	
Results		Activity of each species Γ, but not the number of vibrations in Γ	Number of normal vibrations per species, but not their activity	Vibrations allowed or not; activities follow from IR and Raman spectra and normal vibrations	
Formulae for the irreducbile character	Proper rotations (E, C_n)	$\chi_\mu(k) = 2\cos\psi + 1$ $\qquad$ $\chi_\alpha(k) = 2 + 2\cos\psi + 2\cos2\psi$	$\theta(k) = (m_k - 2)\cdot\chi_\mu(k)$	$\chi_{red}(k) = \chi(\Gamma_1,k)\cdot\chi(\Gamma_2,k)$	Non-degenerate normal vibration: $\chi_{red}(k) = [\chi(\Gamma,k)]^v$ degenerate: see Table 5.12
	Improper rotations (σ,s_n,i)	$\chi_\mu(k) = 2\cos\psi - 1$ $\qquad$ $\chi_\alpha(k) = 2 - 2\cos\psi + 2\cos2\psi$	$\theta(k) = m_k\cdot\chi_\mu(k)$		

Group theoretical reduction formula, character tables

$$Z(\Gamma) = \left[1\bigg/ \sum_k g(k) \right] \cdot \sum_k |g(k)\cdot\chi_{red}(k)\cdot\chi(\Gamma,k)|$$

Reducible character $\qquad$ Irreducible character

From character table for the point group

Results

Type	IR spectra Raman spectra	Normal vibrations	Combinations Overtones
Γ_1 Γ_2 . . . Γ_k	Activity	Number	Decomposition into components Γ_i

mirror plane at a distance $t = T/2$ (cf Fig. 5.3(a)).† The symmetry operation and
the symmetry element are denoted by the letter c.

— *Combination of rotation and translation: screw rotation*   A screw axis of multiplicity p is present when every atom of a molecule A is transferred to an equivalent
atom on molecule A′ by means of a rotation through $\psi = (360/p)°$ followed by a
translation parallel to the axis of rotation, by a distance $t = nT/2$, where $n/p < 1$
(cf Fig. 5.3(b)). The symmetry operation and the symmetry element are denoted
by combining the multiplicity p with the number of translation periods n, for
example 3_1 and 3_2.

† T denotes a full translation period and is connected to the lattice vectors of the unit cell.

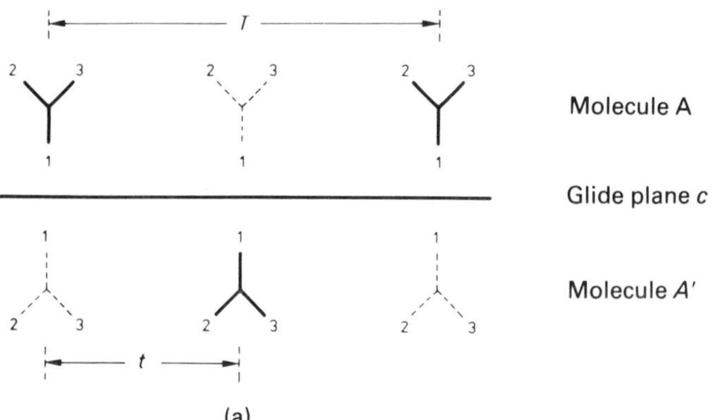

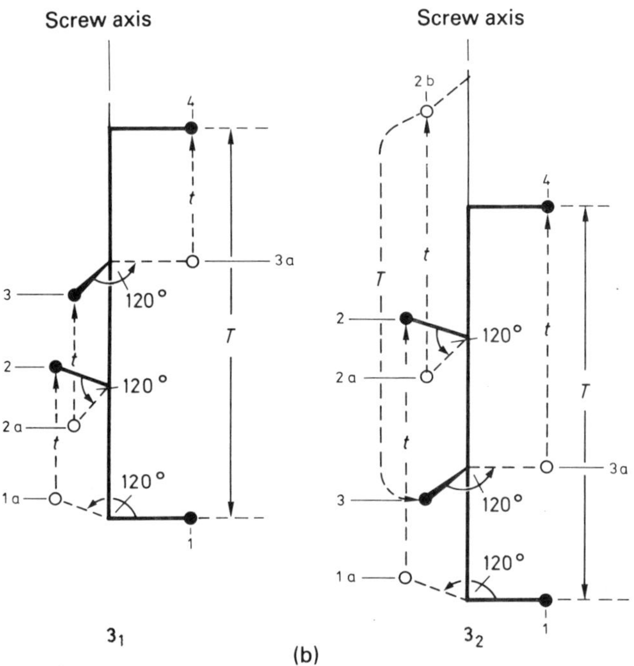

Fig. 5.3 — Open symmetry operations: (a) glide plane; (b) screw axis (example of a screw axis of the order 3).

By combining the open symmetry operations with the point groups which were described in section 5.1.3, the 230 crystallographic space groups are obtained. The group theoretical treatment of solids will be described in Chapter 22.

6

Applications

The determination of the symmetry of a molecule is one of the most important applications of vibrational spectroscopy. This includes the discrimination of isomers, modifications and structure types. This chapter gives some examples of the investigation of isolated molecules (section 6.1) and of molecules in a crystal lattice (section 6.2).

6.1 INVESTIGATIONS OF ISOLATED MOLECULES

By using the methods described in Chapter 5 it is possible to determine the symmetry and the point group of an isolated molecule and therefore the spectroscopic activity of its vibrations. Structural effects, such as co-ordination, which reduce this (i.e. highest) symmetry will become apparent in most cases in the vibrational spectrum.

> *Example 6.1   Fundamental vibrations of the AB_6 octahedron. The case of the AsF_6^- ion*
>
> — The isolated AsF_6^- unit.
> — The distribution of normal vibrations according to the selection rules was given in Example 5.6. This is characterized in more detail in Table 6.1: As a result of inactivity and degeneracy, the infrared spectrum of AsF_6^- shows only two bands, and the Raman spectrum only three bands (one of which is polarized), for the 15 fundamental vibrations of the molecule. Since the octahedron has a centre of symmetry, the mutual exclusion rule has to be obeyed: there are no coinciding bands in the infrared and Raman spectra.

Table 6.1 — Fundamental vibrations of the octahedral AsF_6^- anion modes of vibration according to [21]

Activity	Mode of vibration	Symbol	Degree of degeneracy	Spectrum (schematic)
Raman-active	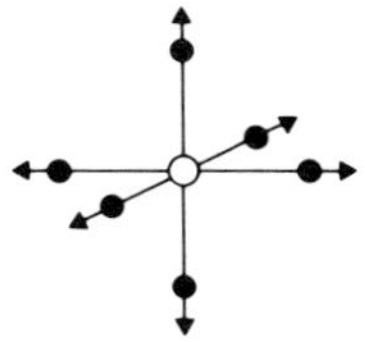	$\nu_s\,(A_{1g})$	(1)	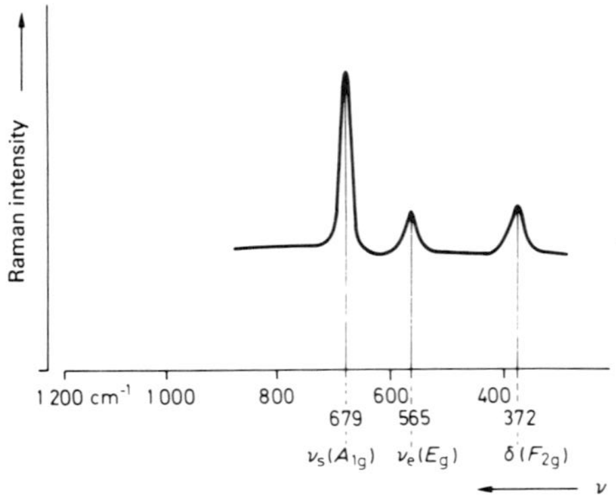
	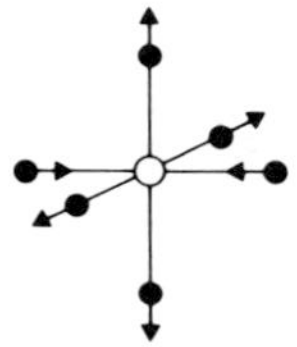	$\nu\,(E_g)$	2	
	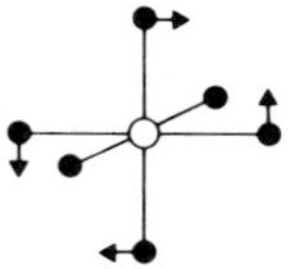	$\delta(F_{2g})$	3	

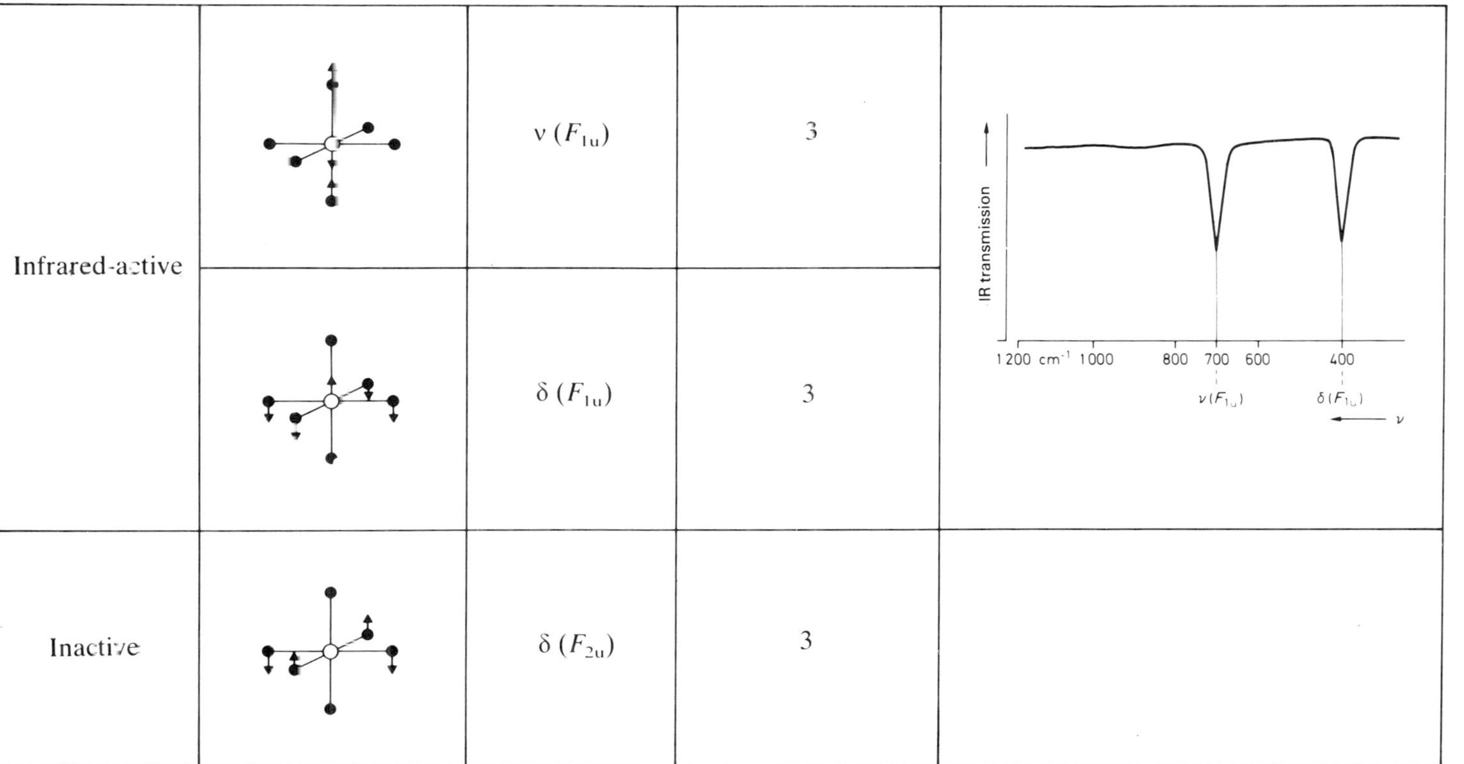

Number of fundamental vibrations: $n = 3 \cdot 7 - 6 = 15$.

— The co-ordinated AsF_6^- unit

The infrared spectrum of the manganese complex $[Mn(NSF_3)_4](AsF_6)_2$ shows in the region of the AsF vibrations (Table 6.2) four bands (720, 703, 674 and

Table 6.2 — A reduction of symmetry caused by co-ordination: fundamental vibrations of a co-ordinated AsF_6^- ion and their activity in C_{4v} symmetry

	Species	Activity IR	Activity Ra	Number of bands	Fundamental vibration
	A_1	+	+	4	$\nu(AsF_{ax,1})$, $\nu(AsF_{ax,2})$, $\nu_s(AsF_4)$ $\delta_s(AsF_4)$ out-of-plane
	A_2	−	−	0	
	B_1	−	+	2	$\nu_{as}(AsF_4)$, $\delta_{as}(AsF_4)$ out-of-plane
	B_2	−	+	1	$\delta_{as}(AsF_4)$
	E	+	+	4	$\nu_e(AsF_4)$, $\delta_e(AsF_4)$, $\delta_e(F_{ax,1}AsF_4)$, $\delta(AsF_4)$ wagging

$590\,\mathrm{cm}^{-1}$) instead of the one band of the octahedral anion ($\nu(F_{1u}) \simeq 700\,\mathrm{cm}^{-1}$). Obviously the degree of degeneracy and inactivity observed in the free AsF_6^- ion must have been reduced by lowering the symmetry [22]. The reason for this is found in the co-ordination of the AsF_6^- ion to the metal atom through one fluorine atom, reducing the symmetry from O_h to C_{4v}. This causes the AsF stretching vibrations $\nu(AsF_{ax,1})$, $\nu(AsF_{ax,2})$, $\nu_s(AsF_4)$, and $\nu(AsF_4)$ to become infrared-active, so that now four bands are observed in the region between 600 and $700\,\mathrm{cm}^{-1}$.

6.1.1 Discrimination between isomers

When the number and kind of atoms in a molecule are known, it is possible to generate the selection rules for the infrared and Raman spectra of all its isomers (sections 5.5.1 and 5.5.2). The distribution of the fundamental vibrations (section 5.5.3) can be determined as well as the activity of combinations (section 5.6.1) and overtones (section 5.6.2). The number of polarized Raman bands (the number of totally symmetric vibrations, cf. Chapter 5) and the number of observable vibrations in both spectra are thus obtained.

The vibrational spectra of isomers differ in most cases in at least one parameter.

Example 6.2   Discriminating between molecules of the composition $Y-X-$
* $X-Y$*

Chain-like molecules of the type $X_2Y_2 = Y-X-X-Y$ can exist in the following point groups: $D_{\infty h}$ (linear), C_{2h}, C_{2v}, and C_2. These systems are distinguishable by the number of infrared active bands and the number of polarized Raman bands (cf. Table 6.3). The groups $D_{\infty h}$ and C_{2h} are centrosymmetric and can be identified by the mutual exclusion principle.

Table 6.3 — The distribution of normal vibrations of a molecule X_2Y_2 as a function of its symmetry

Structure	Point group	Species	Activity		Normal vibrations	$n_{\text{pol}}^{\dagger}$	$n_{\text{IR}}^{\ddagger}$	
			IR	Ra				
Y−X−X−Y	$D_{\infty h}$	Σ_g	−	+	2υ	2	2	Linear
		Σ_u	+	−	υ			
		Π_g	−	+	δ			
		Π_u	+	−	δ			
(X−X angled)	C_{2h}	A_g	−	+	$2\upsilon, \delta$	3	3	Angled
		A_u	+	−	τ			
		B_g	−	+	−			
		B_u	+	−	υ, δ			
(Y−X−X−Y)	C_{2v}	A_1	+	+	$2\upsilon, \delta$	3	5	
		A_2	−	+	τ			
		B_1	+	+	−			
		B_2		+	υ, δ			
(X−X)	C_2	A	+	+	$2\upsilon, \delta, \tau$	4	6	
		B	+	+	υ, δ			

† Number of polarized-Raman bands.
‡ Number of infrared-active vibrations.

The wavenumbers of *trans*- and *cis*-difluorodiazene (F−N=N−F, **3** and **4** respectively) are listed in Table 6.4. The two isomers are easily distinguished from the activity of the symmetric vibrations.

Remark: The presence or absence of bands in characteristic regions, due to the symmetry-related activity of certain vibrations, plays an important role in the determination of molecular symmetry, whereas the total number of active fundamental vibrations by itself is only a relatively weak criterion. This is also true for the number of polarized Raman bands. Also the interpretation of vibrational spectra can be complicated by several factors:

— Weak bands, shoulders or overlapping bands are easily overlooked.
— Combinations and overtones cannot always be distinguished from weak fundamental vibrations. This is especially important in the case of Fermi resonance.
— Complete spectra are required for the interpretations, including the region below the routine infrared range ($< 400 \text{ cm}^{-1}$).

For these reasons the unambiguous determination of symmetry sometimes requires additional information (exact intensities and band positions) as well as the interpretation of other spectra (e.g. NMR).

Table 6.4 — Wavenumbers and activities of the fundamental vibrations of *trans* (**3**) and *cis* (**4**) difluorodiazine (N_2F_2) [23]

C_{2h} (**3**)		C_{2v} (**4**)		Vibration
A_g	$1010^\dagger$	A_1	896	ν_s (NF)
	$1636^\dagger$		1524	ν (NN)
	$592^\dagger$		552	δ_s (NNF)
A_u	360	A_2	$-?-$	τ
B_g	—	B_1	—	—
B_u	989	B_2	952	ν_{as} (NF)
	421		737	δ_{as} (NNF)

† Not infrared-active.

6.1.2 Local symmetry

The interpretation of the vibrational spectra of large molecules often requires a separation into subunits with different *local symmetries* which are equal to or higher than the overall symmetry of the molecule.

Example 6.3   Local symmetries of the F_5As–NSF molecule **5** [22]

Seen as a whole, the adduct F_5As–NSF_3 **5** has C_s symmetry. Seventeen of its 27 fundamental vibrations would then belong to the totally symmetric species A' and the remaining 10 to the species A'' (or 16 in A' and 11 in A'', depending on the position of the AsF_5-group relative to the mirror plane). Both species are infrared- and Raman-active, so that complete coincidence of the two spectra would be expected. If the two components are considered separately, local C_{3v} (NSF_3) and C_{4v} (AsF_5) symmetries (cf. Table 6.5) are found, which leads to the following distribution of the 27 normal vibrations:

$$\Gamma_{vib} = 7A_1 + A_2 + 2B_1 + B_2 + 8E$$

The vibrations of the species B_1 and B_2 are Raman-active, those of the species A_1 and E are infrared- and Raman-active, A_2 is infrared- and Raman-inactive. Only 15 of the bands belonging to normal vibrations should occur in the infrared spectrum, and 7 polarized bands are expected in the Raman spectrum. The separation of the total symmetry of the molecule into two regions of higher local symmetry, which is based on the assumption of independent rotation of the two halves of the molecule about the As–N bond, is consistent with the observed

Table 6.5 — The distribution of the 27 fundamental vibrations of the adduct $F_5As–NSF_3$ to account for local symmetry

Local symmetry	A_1	A_2	B_1	B_2	E
$F_5As–NSF_3$ structure; upper AsF_5 part: C_{4v} lower NSF_3 part: C_{3v} **5**	$\nu\,(AsF_{ax})$ $\nu_s\,(AsF_4)$ $\delta_s\,(AsF_4)$ $\nu\,(NAs)$ — τ — $\nu\,(NS)$ $\nu_s\,(SF_3)$ $\delta_s\,(NSF)$	$\nu_{as}\,(AsF_4)$ $\delta_{as}\,(AsF_4)$		$\delta_{as}\,(AsF_4)$	$\nu\,(AsF_4)$ $\delta\,(AsF_4)$ $\delta\,(F_{ax}AsF_4)$ $\delta\,(NAsF_4)$ $\delta\,(AsNS)$ $\nu\,(SF_3)$ $\delta\,(NSF)$ $\delta\,(FSF)$

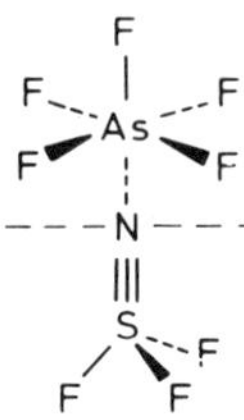

spectra. This model is also supported by calculations, which show that vibrational coupling between the two parts is negligible.

6.2 INVESTIGATIONS OF MOLECULES IN A CRYSTAL LATTICE

All the examples so far have dealt with investigations of isolated molecules where the use of closed symmetry operations is sufficient. In the case of molecules which are part of a crystal lattice, the influence of neighbouring atoms on the symmetry has to be considered as well (cf. open symmetry elements, section 5.8).

6.2.1 Discrimination between structure types

The examples in Chapter 5 have shown that the number of active vibrations decreases as the symmetry of a molecule increases. It is in many cases possible to deduce some information about parts of a crystal structure from the number of bands observed in specific regions of the spectrum.

Example 6.4   The investigation of $A_2^{II}BMO_6$ perovskites (see also section 21.2.3)

According to Kemmler-Sack [24] this class of solids can be separated on the basis of their Raman spectra into three structurally different groups (Fig. 6.1):

(1) Perovskites with undistorted MO_6 octahedra show a maximum of three characteristic bands and fulfill the mutual exclusion rule. The degenerate stretching vibration $\nu(E_g)$ gives rise to very weak bands which are observable only in the case of M = Te or Sb.

(2) The reduced symmetry in the case of distorted MO_6 octahedra affects the number of bands by partly cancelling inactivity and degeneracy. A doublet or triplet is then observed in place of the triply degenerate $\delta(F_{2g})$ deformation.

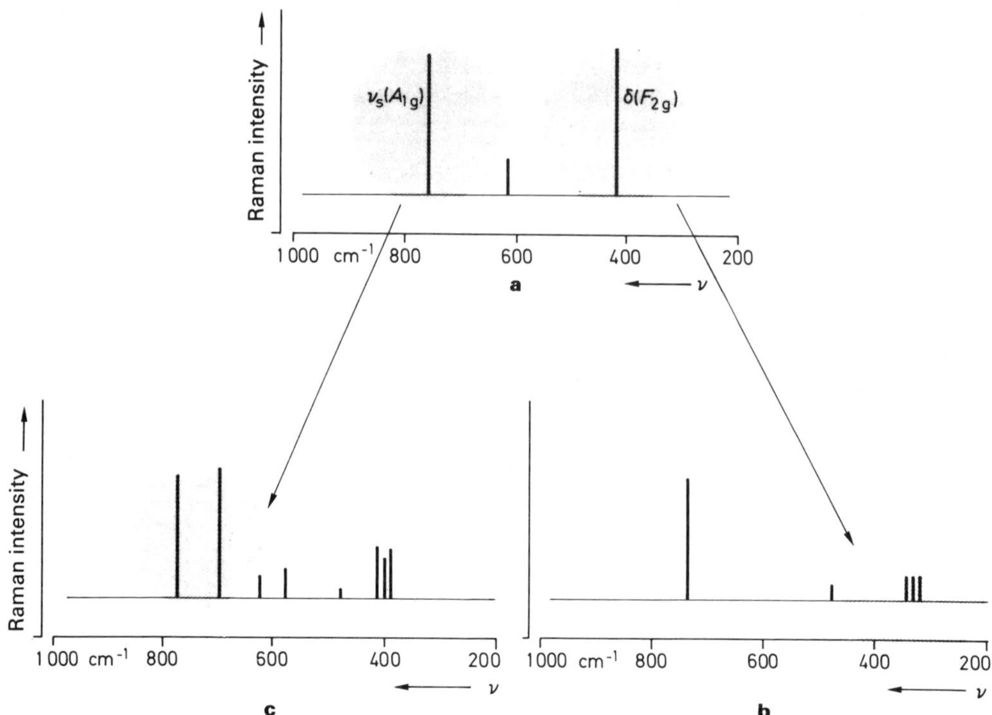

Fig. 6.1 — Schematic representation of the Raman spectra of three types of perovskites: (a) undistorted octahedron, e.g. Ba_2CaTeO_6; (b) equal, distorted octahedra, e.g. Ba_2CaUO_6; (c) other octahedra, e.g. Ba_2ZnTeO_6.

(3) Perovskites with other types of octahedra are easily recognized by the presence of several bands in the region of the non-degenerate $\nu_s(A_{1g})$ vibration.

6.2.2 Discrimination of modifications

Vibrational spectra can be used to differentiate between different modifications of a compound (i.e. different crystal structures of the same composition).

Example 6.5   Calcite and aragonite (see also Example 22.1)
Numerous compounds of the composition MXO_3 crystallize in the aragonite structure if the cation radius is large $(r(M) > 100\,pm)$ and in the calcite structure if the cation radius is small $(r(M) < 100\,pm)$. Calcium carbonate exists in both modifications because $r(Ca^{2+})$ lies on the borderline. They differ in the co-ordination number of the cation: calcite corresponds to a distorted NaCl structure, whereas aragonite is similar to the NiAs type. This has different consequences for the symmetry environment of the carbonate anion (cf. Fig. 6.2 and Table 6.6):

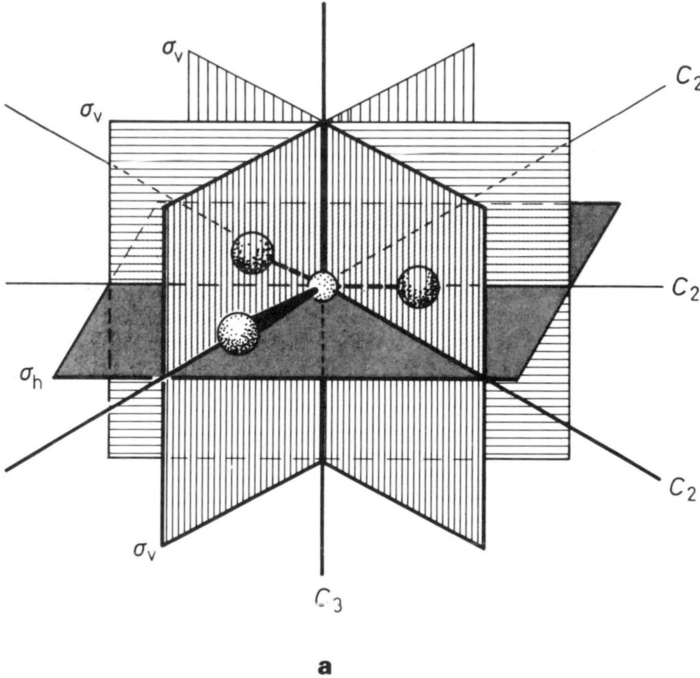

Fig. 6.2 — (a) Symmetry of the isolated carbonate ion;

— The isolated carbonate ion has a threefold axis of rotation (C_3), three perpendicular twofold axes of rotation (C_2) as well as three vertical (σ_v) and one horizontal mirror plane (σ_h). It therefore belongs to the point group D_{3h} (cf. Fig. 6.2(a) and Table 6.6)

— In the *calcite structure* each oxygen atom is co-ordinated to two calcium atoms in such a way that the mirror planes are lost. The symmetry in the crystal lattice is therefore reduced to the point group D_3. However, the selection rules show that the activity of the fundamental vibrations is the same as for the isolated molecule:

The symmetric stretching vibration is infrared-inactive. Two vibrations are doubly degenerate. The infrared spectrum shows only three bands, which belong to the six fundamental vibrations (cf. Fig. 6.2(b)).

— In the *aragonite structure* the symmetry of the carbonate group is reduced even more: the only remaining symmetry element is a mirror plane. Its point group is therefore C_s (cf. Table 6.6). Inactivity and degeneracy are removed completely. $\nu_s(CO_3)$ is now infrared-active and the previously degenerate deformation is split into two bands (the corresponding splitting of the stretching mode is not observed). The group theoretical treatment of the solid (calcite) will be described in Example 22.1.

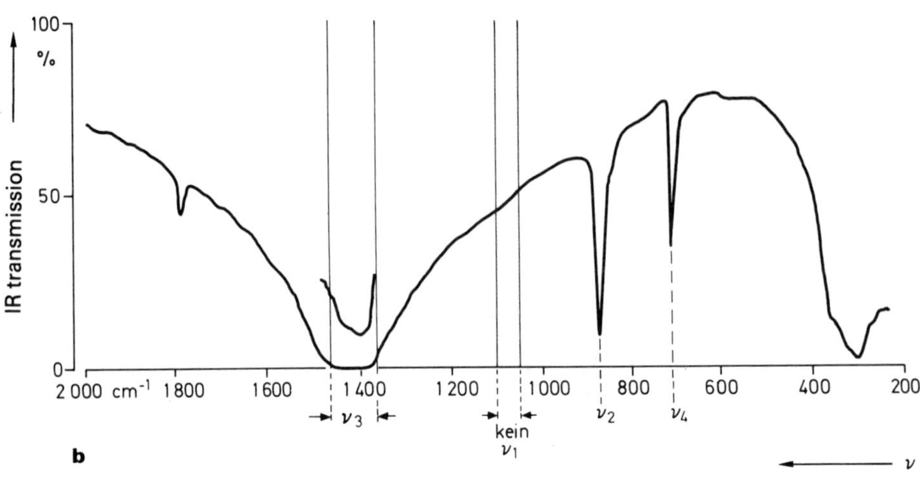

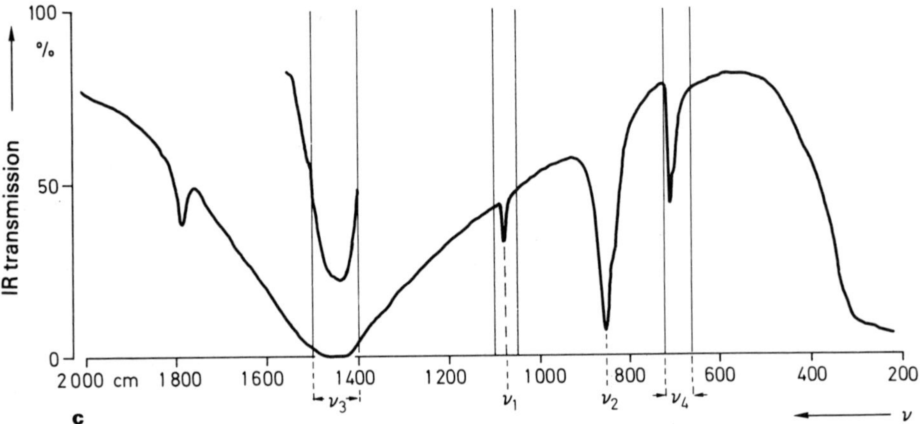

Fig. 6.2 — (b) infrared spectrum of calcite; (c) infrared spectrum of aragonite.

6.3 ORDERING IN MINERALS

Vibrational spectra are sensitive to structure and ordering of solids, as will be illustrated in the following example.

Table 6.6 — Distribution and activity of the fundamental vibrations of the carbonate (CO_3^{2-}) ion

CO_3^{2-} ion	Point group	Species	Activity		Number of bands n
			IR	Ra	
Isolated	D_{3h}	A_1'	$-$	$+$	1
		A_1''	$-$	$-$	0
		A_2'	$-$	$-$	0
		A_2''	$+$	$-$	1
		E'	$+$	$+$	2
		E''	$-$	$+$	0
Calcite	D_3	A_1	$-$	$+$	1
		A_2	$+$	$-$	1
		E	$+$	$+$	2
Argonite	C_s	A'	$+$	$+$	6
		A''	$+$	$+$	6

Example 6.6   Consequences of the ordering in kaolinites [25]
These solids are made up of $[(HO)_4Al_2–Si_2O_5]$ layers. The relevant structural unit is a two-dimensional network of SiO_4 tetrahedra (Si_2O_5 layer). The free oxygen atoms are co-ordinated to the Al^{3+} ions in such a way that they are co-ordinated by six oxygen atoms or hydroxide groups (cf. Fig. 6.3). The infrared

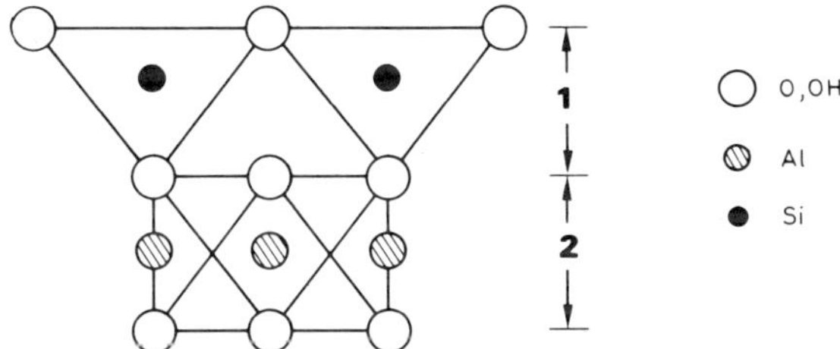

Fig. 6.3 — Schematic representation of the kaolinite structure [26]: **1** Si_2O_5 layer (network of SiO_4 tetrahedra); **2** $Al_2(OH)_4$ layer.

spectrum shows bands in the region of the OH stretch ($v = 4000–3000\,cm^{-1}$) and in the region of the SiO_4 stretch ($v = 1100–400\,cm^{-1}$). Several minerals, which differ in the periodicity and ordering of the layers, can be identified by interpreting the $v(OH)$ region. The classification of kaolinites is shown in Table 6.7 and is based on two parameters: the number n_{OH} of $v(OH)$ bands and the ratio K of the absorbances A of the two most intense bands

$$K = \frac{A(3695\,cm^{-1})}{A(3620\,cm^{-1})}$$

Table 6.7 — Classification of kaolinite minerals by use of infrared spectra

K	n_{OH}	Type
$1.5 \leqslant K \leqslant 1.7$	4	Kaolinite 1 T_1 triclinic
$1.7 \leqslant K$	4	Kaolinite 2 T_2 triclinic
$1.1 \leqslant K \leqslant 1.5$	3	Disorder-type T_d monoclinic
$0.8 \leqslant K \leqslant 1.1$	2	Halloysite 1 M_d
$0.8 \leqslant K \leqslant 1.1$	2	Dickite 2 M

Part IV
Interpretation of molecular spectra by using mathematical tools

The vibrational spectra of molecules are analysed by using the *classical theory of small vibrations* of atoms about their equilibrium position. This extensive and well tested method can provide isotopic frequencies and assignments of unknown vibrations and shows trends in series of homologues. Approximate calculations of bond lengths are possible, and the general insights gained into molecular bonding have helped to solve a number of structural problems. These methods are also applicable to separate molecules in solids.

Chapters 7 and 8 illustrate the simplest cases of force constant calculations for a harmonic oscillator. Chapters 9–12 lay the basis for force constant calculations by using matrices. The methods of application are discussed in Chapters 13–16. Important applications are then summarized in Chapters 17–20.

IV.A
Force constant calculations with the two-mass model

The simple two-mass model allows numerous exact and approximate calculations of force constants of smaller and larger systems, and these give a first impression of the mathematical analysis of vibrational spectra. The two-mass model can be extended to larger molecules by using group theoretical methods in the case of symmetric molecules (cf. sections 7.2 and 7.3) and also by using suitable isotopic substitutions (cf. Chapter 8).

The diatomic model provides a conceptual base for the more extensive methods, since they have many problems in common.

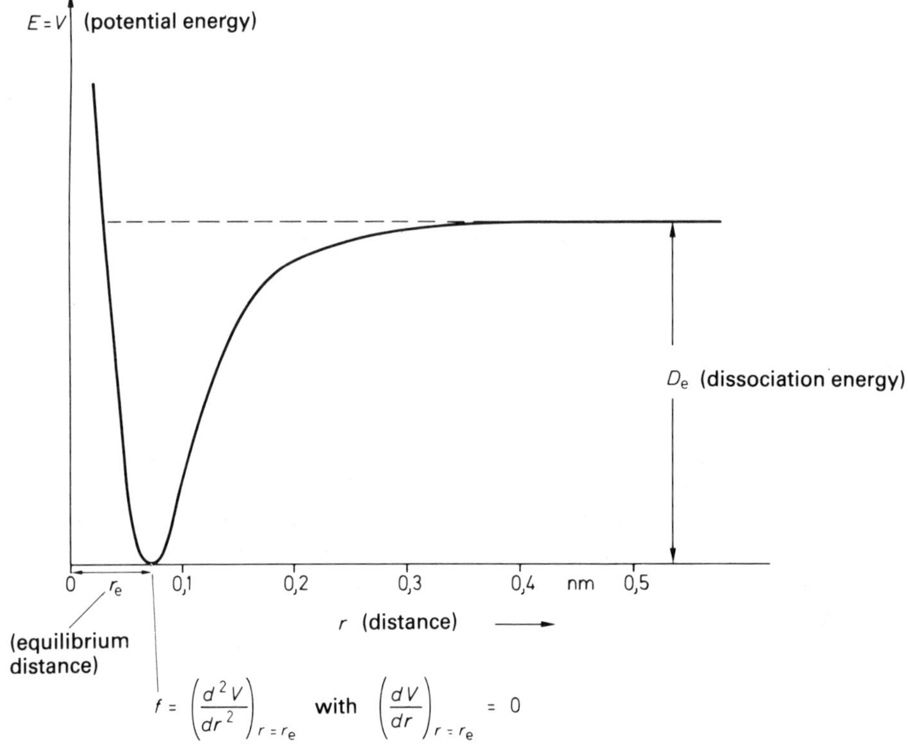

$$f = \left(\frac{d^2 V}{dr^2}\right)_{r=r_e} \quad \text{with} \quad \left(\frac{dV}{dr}\right)_{r=r_e} = 0$$

Fig. 7.1 — The potential energy curve of the H_2 molecule is shown to illustrate the valence force constant $f = f_{XY}$ of a diatomic molecule XY. The valence force constant f is given by the second derivative of the potential energy V over the distance r at the equilibrium distance r_e.

7

Exact and approximate force constant calculations by using the two-mass model

The two-mass model can be used to solve exactly for the force constants for diatomic molecules, highly symmetrical molecules such as XY_2 ($D_{\infty h}$), X_4 (T_d), etc. and parts of other symmetrical systems (e.g. XY_6 (O_h)). It also allows some force constants of more complex, less symmetrical molecules to be estimated with reasonable accuracy.

7.1 DIATOMIC MOLECULES

The *force constant* $f(XY)$ of a diatomic system XY with $C_{\infty v}$ symmetry can be calculated from the following equation [27,28] (cf. Table 1.4):

$$f(XY) = \frac{C v_{XY}^2}{\mu_X - \mu_Y} \tag{7.1.1}$$

in N/cm (Newton/centimetre) [earlier mdyne/Å], where v_{XY} is given in cm^{-1} and the reciprocal masses μ_X and μ_Y are given in the atomic scale relative to ^{12}C, which gives rise to the conversion factor $c = 5.89146 \times 10^{-7}$.

> *Example 7.1* $^{14}N\,^{16}O$
> From $v(NO) = 1876\,\text{cm}^{-1}$ (IR) [29] and $m_N = 14.003$ and $m_O = 15.995$ the valence force constant for the N$-$O bond, $f(NO)$, is found to be 15.48 N/cm [30].

7.2 TRI- AND TETRA-ATOMIC SYSTEMS

The group theoretical analysis for some highly symmetric molecules leads to a number of separable problems which can treated according to the two-mass model. Among these are the chemically important systems XY_2 ($D_{\infty h}$) (e.g. CO_2, CS_2, N_3^-) and X_4 (T_d) (e.g. P_4, As_4).

Example 7.2 $XY_2 (\boldsymbol{D}_{\infty h})$
The application of group theory leads to the following set of equations for the valence force constant f_{XY}, the valence–valence coupling force constant $f_{XY/XY}$ and the deformation force constant $f_{YXY} = f_\alpha$ [31] (cf. Fig. 1.5):

$$(f_{XY} + f_{XY/XY})\, \mu_Y = cv_s^2 \tag{7.2.1}$$

$$(f_{XY} - f_{XY/XY})\, (\mu_Y + \mu_X) = cv_{as}^2 \tag{7.2.2}$$

$$f_{YXY}\, 2(\mu_Y + 2\mu_X) = cv_\alpha^2 \tag{7.23}$$

The N_3^- ion
For the N_3^- ion in crystalline, KN_3, the following frequencies have been observed: $v_s = 1344\,\text{cm}^{-1}$, $v_{as} \doteq 2041\,\text{cm}^{-1}$ and $v_\alpha = 645\,\text{cm}^{-1}$ by Raman and infrared spectroscopy respectively [32]†. The mass $m_N = 14.01$ will be used. The calculations lead to the following force constants for the azide ion:

$$f_{NN} = 13.18\,\text{N/cm}\ , \qquad f_{NN/NN} = 1.72\,\text{N/cm}$$

and

$$f_{NNN} = f_\alpha = 0.57\,\text{N/cm}\ [33]$$

7.3   INDIVIDUAL FORCE CONSTANTS OF POLYATOM SYSTEMS

In numerous polyatom, symmetric molecules there appear simple linear equations for some force constants as well as eigenvalue problems of higher order. In the following we will give one example each for the stretching vibrations of 4-, 5-, 6-, and 7-atom systems, and for a deformation, a torsion and an out-of-plane bending vibration.

Valence of the planar XY_3 system (D_{3h})
The symmetry force constant $F_{11}\,(A_1')$ of the totally symmetric (A_1') stretching vibration is composed of the valence force constant f_{XY} and the two valence–valence coupling constants $f_{XY/XY}$:

$$F_{11}(A_1') = F_{XY} + 2f_{XY/XY} = m_Y cv^2(A_1') \tag{7.3.1}$$

(cf. Fig. 10.2, Eq. (12.4) and Example 12.2).

Example 7.3 SO_3
With $v(A_1') = 1065\,\text{cm}^{-1}$ (Ra) [34] and the atomic mass $m_Y = m_O = 16$, the symmetry force constant $F_{11}(A_1')$ is found to be 10.69 N/cm.

---

† From here onwards we will follow the practice of using the units of cm^{-1} for vibrational frequencies, even though it is strictly speaking incorrect.

Valence of the tetrahedral XY_4 ***system*** ***(T_d)***
Similarly the symmetry force constant $F_{11}(A_1)$ of the totally symmetric mode A_1 is given by Eq. (12.4) (cf. Fig 5.1 3):

$$F_{11}(A_1) = f_{XY} + 3f_{XY/XY} = m_Y, \; cv^2(A_1) \tag{7.3.2}$$

where three coupling constants, to the three other equivalent bonds, have to be added to the valence force constant (cf. Example 12.2).

Example 7.4   SiF$_4$
Based on the frequency $v(A_1) = 801 \, \text{cm}^{-1}$ (Ra) [35] and the atomic mass $m_Y = m_E = 19$ the symmetry force constant of $F_{11}(A_1)$ is found to be 7.18 N/cm for this 5-atom system.

Valences of the trigonal bipyramid XY_5 ***(D_{3h})***
This system consists of two axial valences XY_{ax} along the YXY axes and three equatorial valences XY_{eq}. The coupling between the two symmetric A_1' modes of the axial and equatorial valences is zero. The axial and equatorial symmetry valence force constants can therefore be calculated from

$$F_{11,ax}(A_1') = f_{ax} + 2f_{ax/ax} = m_Y cv_{ax}^2(A_1') \tag{7.3.3}$$

and

$$F_{22,eq}(A_1') = f_{eq} + 2f_{eq/eq} = m_Y cv_{eq}^2(A_1') \tag{7.3.4}$$

(cf. Eq. (12.4))

Example 7.5   SbCl$_5$
Based on the frequencies $v_{ax}(A_1') = 307 \, \text{cm}^{-1}$ and $v_{eq}(A_1') = 357 \, \text{cm}^{-1}$ (Ra) [36] and the atomic mass $m_Y = m_{Cl} = 35.45$, the axial symmetry force constant $F_{11,ax}(A_1')$ is equal to 1.97 N/cm, and for the equatorial symmetry force constant $F_{22,eq}(A_1')$, 2.66 N/cm.

Valences of the octahedral XY_6 **system** **(O_h)**
The symmetry force constant $F_{11}(A_{1g})$ for the totally symmetric vibration $v_1(A_{1g}) = v_1(XY)$ of the XY valence is given by

$$F_{11}(A_{1g}) - f_{XY} + 4f_{XY/XY}(90°) + f_{XY/XY}(180°) = \frac{cv_1^2(A_{1g})}{\mu_Y} \tag{7.3.5}$$

and symmetry force constant for the doubly degenerate vibration $v_2(E_g) = v_2(XY)$ is given by [37]

$$F_{22}(E_g) = f_{XY} - 2f_{XY/XY}(90°) + f_{XY/XY}(180°) = \frac{cv_2^2(E_g)}{\mu_Y} \qquad (7.3.6)$$

where f_{XY} is the valence force constant for the valence XY, $f_{XY/XY}(90°)$ is the coupling constant between two perpendicular valences and $f_{XY/XY}(180°)$ is the coupling constant between two valences that lie on the same line. The coupling constant $f_{XY/XY}(90°)$ can be obtained by combining the two equations.

$$f_{XY/XY}(90°) = \frac{F_{11}(A_{1g}) - F_{22}(E_g)}{6} \qquad (7.3.7)$$

Example 7.6 SF$_6$
With the frequencies $v_1(A_{1g}) = 774\,\text{cm}^{-1}$ and $v_2(E_g) = 642\,\text{cm}^{-2}(\text{Ra})$ [38] and the mass $m_F = 18.998$ the the symmetry force constants obtained are

$$F_{11}(A_{1g}) = 6.71\,\text{N/cm}$$

$$F_{22}(E_g) = 4.61\,\text{N/cm}$$

and the coupling constant is

$$f_{XY/XY}(90°) = 0.35\,\text{N/cm}$$

Some applications. Individual valence force constants are used in the computation of isotopic frequencies, as discussed in Chapter 17. The computation of force constants in homologous or other series of compounds can help to clarify bonding and structural properties, as can be seen from the compilation of force constants of frequently occurring chemical bonds which has been included in the Appendix.

Large coupling constants point to strong non-bonding interactions between atoms in a molecule [39].

Deformation force constant of the linear XYZ **system** ($C_{\infty v}$)
The deformation force constant f_α of the Π mode is (cf. Fig. 10.1)

$$f_\alpha = \frac{cv_\alpha^2}{(\mu_X v^{-1} + \mu_Z v + \mu_Y(v^{-1} + 2 + v))} \qquad (7.3.8)$$

where $v = r_{XY}/r_{XZ}$.

Example 7.7 Deformation force constant of ClCN
From the deformation frequency $v_\alpha = 380\,\text{cm}^{-1}$ (IR) [40] and the ratio $v = 1.631/1.159 = 1.407$ [41] the deformation force constant calculated is

$$f_\alpha(\text{ClCN}) = 0.18\,\text{N/cm}$$

***Torsion force constant of the planar* YXXY *system* ($C_{\infty v}$)**
For the torsion vibration ν_τ of the species A_u [42, 43]

$$f_r = \frac{c\omega_\tau^2}{2\sin^{-2}\alpha(\mu_X + \mu_Y)} \tag{7.3.9}$$

Example 7.8   Torsion force constant of N_2F_2 (C_{2h})
The torsion frequency of the *trans* form of N_2F_2, $\nu_\tau = 364\,\text{cm}^{-1}$ (IR) [44], and the angle $\alpha = 115°$ [45] give rise to a torsion force constant

$$f_t(N_2F_2(C_{2h})) = 0.38\,\text{N/cm}$$

***Out-of-plane constant* f_γ *of the planar system* XY_3 (D_{3h})**
From the out-of-plane vibration ν_γ of the species A_2'' it follows that [46,47] (cf. Table 1.4 and Fig. 10.2)

$$f_\gamma = \frac{c\nu_\gamma^2}{3(\mu_Y + \mu_X)}$$

Example 7.9 $^{10}BF_3$
From the measured frequency $\nu_\tau = 718\,\text{cm}^{-1}$ (IR) for the out-of-plane motion [48], the force constant is found to be

$$f_\gamma(BF_3) = 0.29\,\text{N/cm}$$

Comparison and significance of the shape-restoring force constants
Examples 7.7 to 7.9 show that the shape-restoring deformation, torsion and out-of-plane deformation force constants are about an order of magnitude lower than the valence force constants. This means that the deformation of a molecule requires less energy than the change of bond length.

High-pressure studies have confirmed the shape-restoring function of the deformation force constant (cf. section 23.6).

In homologous series, numerous trends appear in the deformation force constants [41].

7.4   APPROXIMATE COMPUTATION OF FORCE CONSTANTS OF POLYATOMIC MOLECULES

In many cases, estimates of the valence force constants of polyatomic molecules from the two-mass model can be sufficient. Such estimates can yield sufficiently accurate values from *characteristic vibrations* for, for example, $H-C$ and $H-O$ bonds.

Table 7.1 contains data for four triatomic molecules for which the force constants

Table 7.1 — Valence frequencies and valence force constants of some selected triatomic molecules computed by using the two-mass model. Valence force constants obtained from complete data sets are shown in parentheses for comparison

Molecule	Vibrational frequencies (IR) [49] cm^{-1}		Valence force constants	
XYZ	ν_{XY}	ν_{YZ}	F_{XY}	f_{YZ}
HCN ($C_{\infty v}$)	3311	2097	6.01 (5.81) [50]	16.75 (18.07) [50]
ClCN ($C_{\infty v}$)	714	2219	2.69 (4.65)[†]	18.76 (18.33)[†]
ClNO (C_s)	332	1800	0.65 (2.39) [51]	14.25 (15.18) [51]
SeCS ($C_{\infty v}$)	506	1435	1.57 (5.97) [52]	10.60 (7.6) [52]

† cf. Example 14.1.

of the XY and YZ valences have been computed with the two-mass model by using the characteristic vibrational frequencies. Data from complete vibrational analyses have been included for comparison. HCN has a high H−C frequency and a dominating f_{HC} force constant. In the case of ClCN the CN frequency is the higher one and and f_{CN} is correspondingly stronger. The valence force constant f_{ClC} is almost halved according to the two-mass model and would be useless for estimates. The comparison of the two-mass model with the complete data set shows a reduction of the large f_{CN} force constant in favour of the smaller f_{ClC} force constant in the latter case.

For ClNO, the opposite behaviour is observed for the large f_{NO} force constant, which increases from 14.25 for the decoupled case to 15.18 N/cm for the complete data set. The force constant f_{ClN} in the two-mass approximation is at one third of its exact value and therefore useless. The approximations in the case of SeCS are again very poor.

Finally, the following are some results for the N−N bond in a 14-atom azide and a 29-atom diazoalkane [51].

$$N-NNC_6H_5: \nu_{NN} = 2096\,\text{cm}^{-1}\ (IR)\ [53]\ ;$$

$$f_{NN} = 18.12\ (16.3)\,\text{N/cm}$$

(terminal N−N bond)

$$NNC(Sn(CH_3)_3): \nu_{NN} = 2008\,\text{cm}^{-1}\ (IR)\ [54]\ ;$$

$$f_{NN} = 16.63\ (16.48)\,\text{N/cm}$$

The difference between the result from the two-mass model and the exact value is about 10% in the case of the 14-atom $N-NNC_6H_5$ and less than 1% in the case of the 29-atom $NNC(Sn(CH_3)_3)_2$ (cf. Example 8.1).

It should be noted that the results obtained for large molecules with the two-mass model are often taken as the starting point for other methods which will be discussed in detail in Chapters 14 and 16. This method largely reduces the discrepancies observed for HCN, ClCN and partly also for ClNO, but not for SeCS, which demonstrates the requirement for other approaches [55].

Note: Because of the explosion hazard associated with azides and diazoalkanes, infrared spectroscopy is the preferred method for the investigation of these compounds (cf. section 23.3).

8

The exact calculation of individual force constants for symmetrical polyatomic molecules by specific isotope substitution

Some valence force constants f_{ZX} of large molecules of the kind $ZXYR_n$ and ZXR_n with C_{nv} symmetry can be determined if the vibrational frequencies of the isotopic compounds $Z'XYR_n$ and $Z'XR_n$ are known [55–57]:

$$f_{ZX} = \frac{5.89146 \times 10^{-7}}{\mu_Z - \mu_{Z'}} \sum_{i=1}^{p} (v_i^2 - v_i'^2) \tag{8.1}$$

where

$\quad f_{ZX} \qquad$ = valence force constant for the bond ZX in N/cm,

$\quad \mu_Z \qquad$ = m_Z^{-1} = reciprocal mass of the atom Z,

$\quad \mu_{Z'} \qquad$ = $m'_Z{}^{-1}$ = reciprocal mass of the atom Z' (atomic mass units),

$\quad v_i \qquad$ = ith vibrational wave number (cm^{-1}) of the molecule $ZXYR_n$ or ZXR_n for totally symmetric species ($i = 1, 2, \ldots, p$)

$\quad v_i' \qquad$ = ith vibrational wave number (cm^{-1}) of the isotopic molecules $Z'XYR_n$ or $Z'XR_n$ for totally symmetric species ($i = 1, 2, \ldots, p$).

The valence force constant f_{ZX} is now only a function of the atomic masses m_Z and $m_{Z'}$ and the vibrational wave numbers v_i and v_i' of the totally symmetric modes of the two compounds.

Example 8.1   The force constant f_{NN} for $NNCH_2$ (C_{2v})
With the masses $m_N = 14.003$ and $m_N = 15.000$ together with the vibrational wave numbers listed below, the valence force constant is calculated to be $f_{NN} = 18.09$ N/cm for diazomethane [53].

Diazomethane	ν (NN)	ν (CN)
$^{14}N_2CH_2$	2097	1136
$^{15}N^{14}NCH_2$	2075	1112
f_i N/cm	11.39	6.70
f_{NN} N/cm	18.09	

The valence force constant f_{NN} for $NNC(Sn(CH_3)_3)_2$ (16.48 N/cm) shows the influence of the substituents on the N−N bonds; in this case the π-bond is weakened.

IV.B
The basics of matrix algebra — energy matrices — general valence force fields — symmetric molecules

The computational analysis and chemical interpretation of the vibrational spectra of polyatomic molecules require some new concepts and models:

The numerous parameters, co-ordinates and constants can be neatly grouped by using *matrices*. To successfully apply these we need to know the rules of matrix algebra (Chapter 9).

The physicochemical interpretation of the vibrations of polyatomic molecules is based on the *general valence force field*. The valence forces and their coupling are only in a few cases sufficient to maintain interatomic distances. Normally deformation forces and their couplings have to be included to describe all vibrations (Chapter 10).

The chemical valence force model, using internal co-ordinates (bond lengths and angles), leads to the determination of *potential energy constants*, which in addition to masses also include distances and angles (Chapter 11).

Symmetrical molecules have to be analysed *group theoretically* to allow the assignment of vibrations to separate uncoupled parts of a molecule (Chapter 12).

9

Basic concepts and definitions of matrix algebra

Matrix algebra [58] allows a clear and concise description of spectroscopic processes [59]. It also facilitates the programming of mathematical methods for the computer-assisted analysis of spectroscopic data.

9.1   MATRICES

Large amounts of numeric or other data are often collected in tables where each item is assigned to a horizontal and a vertical reference. Tables are ordered rectangular collections of data. They can be square and may consist of a single row or column only. Tables that are ordered sets of elements are called matrices in mathematics. A rectangular matrix $\mathbf{A}$ of the elements a_{ik} is defined as follows:

$$\mathbf{A} = (a_{ik}) = \begin{bmatrix} a_{11} & a_{12} & \cdots & a_{1n} \\ a_{21} & a_{22} & \cdots & a_{2n} \\ & \vdots & \vdots & \vdots \\ a_{m1} & a_{m2} & \cdots & a_{mn} \end{bmatrix} \tag{9.1.1.}$$

for a_{ik} with $i = 1, 2, \ldots, m;\ k = 1, 2, \ldots, n.$

A rectangular matrix consists of m rows and n columns and has $m \times n$ coefficients or elements. In vibrational spectroscopy a rectangular matrix appears in the transition from the $3N$ Cartesian co-ordinates to the $3N - 6$ or 5 internal co-ordinates (cf. Eq. (11.1)).

When a matrix has an equal number of rows m and columns n it is called a square matrix of order n, and it has n^2 elements or coefficients a_{ik}.

A matrix $\mathbf{A}$ where

$$\mathbf{A} = (a_{ik}) = (a_{ki}) \tag{9.1.2}'$$

is called a symmetrical square matrix. The number of different elements is then reduced to $n(n+1)/2$.

The matrices of the potential energy (force constant matrices) and of the kinetic energy (so-called **G**-matrices) are square and symmetrical (cf. Eqs. (10.1), (11.9), (11.9a), etc.).

Another case of a square, symmetrical matrix is the diagonal matrix **D**, where only the elements on the main diagonal are not equal to zero.

$$\mathbf{D} = (d_{ii}) = \begin{bmatrix} d_{11} & & & \\ & d_{22} & & \\ & & \ddots & \\ & & & d_{nn} \end{bmatrix} \qquad (9.1.3)$$

where $d_{ii} \neq 0$ and $d_{ik} = 0$ for $i \neq k$.

The matrices of vibrational frequencies **N** and of eigenvalues Λ are diagonal matrices (cf. Example 13.1).

The unit matrix **E** is a special case of the diagonal matrix in which all the diagonal elements are equal to one. In matrix algebra it is equivalent to the figure 1 in ordinary algebra.

An important concept for many applications is that of the regular or non-singular square matrix **A**, which is a square matrix for which the determinant of **A** does not vanish [61], i.e.

$$\det \mathbf{A} \neq 0 \qquad (9.1.4)$$

This is obviously true for the matrices of frequency **N** and of eigenvalues Λ. For physical reasons this is also true for the force constant matrix **F** and the **G**-matrix of the kinetic energy of a vibrating molecule. (Non-square matrices are neither regular nor singular.)

Finally we mention the column matrix, which consists of a single column with m elements. For example the kth column matrix in **A** is

$$\mathbf{a}_k = \begin{bmatrix} a_{1k} \\ a_{2k} \\ \vdots \\ a_{mk} \end{bmatrix} \qquad (9.1.5)$$

with $k = 1, 2, \ldots, n$.

The ith row matrix consists then of one single row of n elements:

$$\mathbf{a}_i = (a_{i1}, a_{i2}, \ldots, a_{in})$$

with

$$i = 1, 2, \ldots, m. \tag{9.1.6}$$

It is important to note that rectangular matrices can be built up from row and column matrices. Column matrices represent vectors; this facilitates access to all co-ordinate systems (cf. Eq. (11.1)).

Rectangular and square matrices are denoted by bold, upper case letters, column and row matrices by bold, lower case letters.

9.2 MATRIX OPERATIONS

The concept of a matrix as a mathematical expression for tables derives its deeper meaning for a number of operations.

Transposed or mirrored matrices

If the rows and columns of a rectangular matrix $\mathbf{A} = (a_{ik})$ are exchanged, the transposed matrix is obtained:

$$\mathbf{A}' = (a_{ki}) = (a'_{ik}) \tag{9.2.1}$$

If a matrix $\mathbf{A}$ is transposed twice, the original matrix $\mathbf{A}$ is obtained, i.e.

$$(\mathbf{A}')' = \mathbf{A} \tag{9.2.2}$$

This operation is required for numerous theoretical and applied studies. We find, for example, that for square, symmetric matrices of the force constants ($\mathbf{F}$) and the kinetic energy ($\mathbf{G}$) (cf. Chapters 10 and 11) generally that

$$\mathbf{A} = \mathbf{A}' \qquad \text{or} \qquad a_{ik} = a_{ki} \tag{9.2.3}$$

Addition and subtraction of matrices

If two matrices $\mathbf{A}$ and $\mathbf{B}$ have the same number of rows and columns then the new matrix $\mathbf{C}$ is formed by addition (subtraction) of $\mathbf{A}$ and $\mathbf{B}$ as follows:

$$\mathbf{C} = \mathbf{A} \pm \mathbf{B} = (c_{ik}) = (a_{ik}) \pm (b_{ik}) \tag{9.2.4}$$

where

$$c_{ik} = a_{ik} \pm b_{ik}$$

In vibrational spectroscopy, this operation is used in the iteration method for the calculation of the force constant matrices.

Matrix multiplication

The most important matrix operation is the multiplication of two rectangular matrices **A** and **B**. Matrix multiplication is the essential piece of matrix algebra, which owes its practical importance to this operation. It is defined as follows:

$$\mathbf{A} \cdot \mathbf{B} = (a_{ir}) \cdot (b_{rk}) = (c_{ik}) = \mathbf{C} \tag{9.2.5}$$

where

$$c_{ik} = \sum_{r=1}^{n} a_{ir} \times b_{rk} \tag{9.2.6}$$

$$for\ i = 1, 2, \ldots, m,$$
$$for\ k = 1, 2, \ldots, p,$$
$$for\ r = 1, 2, \ldots, n.$$

From this follows the condition that the number of rows of the matrix **B** must be equal to the number of columns of the matrix **A**. Often this condition is not fulfilled, as in the case **B** · **A**, which means that this operation does not commute. This distinguishes matrix multiplication from the multiplication of real numbers, which is commutative (i.e. it is independent of the order of multiplication).

Some applications include, the representation of systems of linear equations, co-ordinate transformations (cf. Eq. (11.1)) and transformation of energy matrices from one co-ordinate system to another (cf. Eq. (11.2)).

Inverse matrices

The inverse matrix is defined for a non-singular square matrix **A** as

$$\mathbf{A} \cdot \mathbf{A}^{-1} = \mathbf{A}^{-1} \cdot \mathbf{A} = \mathbf{E}\ \text{(unit matrix)} \tag{9.2.7}$$

The inverse matrix $\mathbf{A}^{-1}$ is also non-singular. If the matrix **A** is symmetric then the inverse matrix is also symmetric [62]. It is not used very often in vibrational analysis.

9.3 APPLICATIONS TO VIBRATIONAL SPECTROSCOPY

For the vibrations of atoms about their equilibrium positions in a molecule we use the model of small vibrations, which can be described elegantly with matrices. For the extensive deductions and definitions we refer the reader to the literature [63,64]. We will only use a few of the results.

For reasons of stability it follows from the harmonic approximation that all main diagonal elements and all main segment determinants of the energy matrices **G** and the force constant matrices **F** must be positive, e.g.:

$$f_{ii} > 0 \tag{9.3.1}$$

$$\begin{vmatrix} f_{11} & f_{12} \\ f_{12} & f_{22} \end{vmatrix} = f_{11} \times f_{22} - f_{12}^2 > 0 \tag{9.3.2}$$

$$\det \mathbf{F} > 0 \tag{9.3.3}$$

Positive force constants f_{ii} along the main diagonal correspond to the bonding and shape-restoring forces that are acting in the molecule (cf. Chapter 10). The expression for the main segment determinants restricts the ranges for the coupling constants.

The analytical solution of the equation of motion of the atoms about their equilibrium position has led, beyond the stability conditions, to a computational analysis of spectroscopic measurements. As in the case of the two-mass model there exists a relationship between the n vibrational frequencies v_i, the N atomic masses and the force constants f_{ik} ($i, k = 1, 2, \ldots, n$), which in the case of internal co-ordinates also includes angles and distances as well as ratios of distances. This relation is the well known secular equation (for arbitrary co-ordinates):

$$\det (\mathbf{G} \cdot \mathbf{F} - \lambda \mathbf{E}) = 0 \tag{9.3.4}$$

where $\mathbf{E}$ is the unit matrix and λ_i are the n eigenvalues which correspond to the squares of the n vibrational frequencies.

Unlike the two-mass model, the number of unknown force constants, which is $n(n + 1)$ for orders $n > 1$, is now greater than the number of algebraic equations n, which correspond to the n vibrational frequencies. In the case of order $n = 2$ there are only two equations for the two frequencies to determine three force constants f_{11}, f_{22} and f_{12} (cf. Chapter 13, Eqs. (13.1) and (13.2)). Numerous mathematical and physicochemical methods have been developed to overcome this underdetermination. Some of these will be treated in more detail in Chapters 13–16.

The determinant of the force constant matrix $\mathbf{F}$ can be calculated from

$$\det \mathbf{F} = \det \Lambda / \det \mathbf{G} \tag{9.3.5}$$

where Λ is the diagonal matrix of the n eigenvalues $\lambda_i, i = 1, 2, \ldots, n$ (Eqs. (13.20)). This equation has several consequences when the frequencies of isotopic compounds are included.

Since the force constants of isotopic compounds are equal the following equation is valid:

$$\det \mathbf{F} = \det \Lambda / \det \mathbf{G} = \det \Lambda_{iso} / \det \mathbf{G}_{iso} \tag{9.3.6}$$

This reduces the number of algebraically independent equations by one. Eq. (9.3.6) can under certain circumstances be used to calculate geometrical parameters such as bond angles (cf. section 23.6). It is often used to estimate isotopic frequencies (cf. Example 17.1).

10

Construction of a force constant matrix, as matrix of the potential energy, according to the general valence force model

For systems with more than two atoms there exist n force constants, i.e. constants of the potential energy of the molecule, for n fundamental vibrations, which can in only a few cases be interpreted as valence force constants in symmetry co-ordinates (cf. Chapter 12). One of these is the tetrahedral X_4 (T_d) molecule, where 4 atoms are connected via 6 bonds. Here $n = 3 \times 4$ atoms $- 6$ degrees of freedom $= 6$ vibrations are completely described by 6 valence force constants. In addition there are $n(n-1)/2$ coupling force constants between the 6 valence force constants, i.e. 15 valence–valence coupling constants. Two examples are P_4 and As_4.

As in the case of the diatomic molecule the valence force constants are defined by the parameters of only two atoms such as masses and vibrational frequencies.

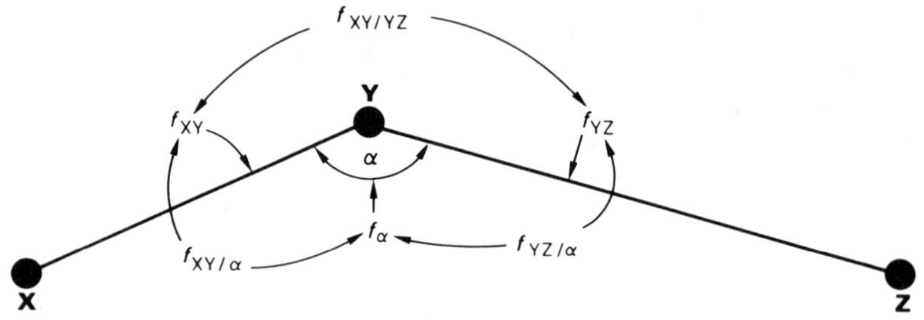

Fig. 10.1 — The two valence force constants f_{XY} and f_{YZ}, the deformation force constant f_α and the three coupling constants $f_{XY/YZ}$, $f_{XY/\alpha}$ and $f_{YZ/\alpha}$.

For most molecules the number of bonds is smaller than the number of fundamental vibrations. These are complemented by angle-restoring force constants, which are known as deformation force constants. The definition of a deformation force constant requires three atoms of the molecule.

The whole force field of a molecule can be represented very clearly in the form of force constant matrices, as the following example of a triatomic XYZ molecule with C_s symmetry illustrates (Fig. 10.1).

The force constant matrix is of order 3 because $n = 3 \times 3 - 6 = 3$.

$$
F(XYZ(C_s)) = \begin{bmatrix} f_{XY} & f_{XY/YZ} & F_{XY/\alpha} \\ f_{XY/YZ} & f_{YZ} & f_{YZ/\alpha} \\ f_{XY/\alpha} & f_{YZ/\alpha} & f_\alpha \end{bmatrix} \cdot \begin{matrix} XY \\ YZ \\ \alpha \end{matrix} \qquad (10.1)
$$

In planar molecules the angles are often interdependent, as the planar system XY_3 (D_{3h}) shown in Fig. 10.2 illustrates. In this case only two deformation constants can

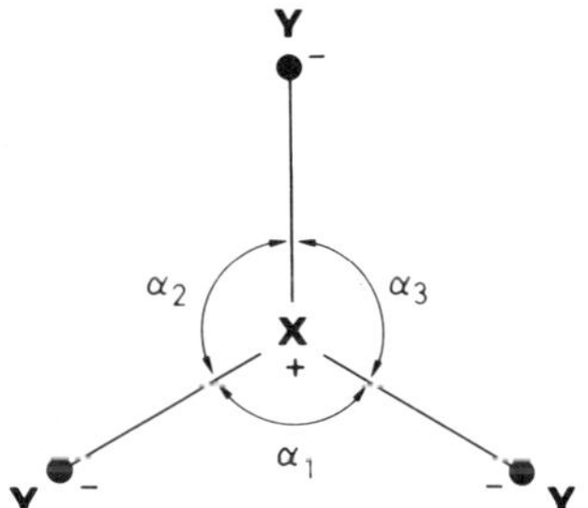

Fig. 10.2 — In addition to the three valence force constants f_{XY} and the deformation constants f_{α_1} and f_{α_2} of the system XY_3 (D_{3h}), there now appears the out-of-plane force constant f_γ. This vibration consists of the movement of the central X atom above the plane of the paper (indicated by a $+$) while the three Y atoms move below the plane (indicated by a $-$).

be added to the three valence force constants, since the three angles in the plane add up to 360°. This leads to a new kind of deformation constant, namely the out-of-plane force constant, which corresponds to the out-of-plane deformation of the system and includes four atoms. This completes the required set of $n = 3 \times 4 - 6 = 6$ force constants on the main diagonal.

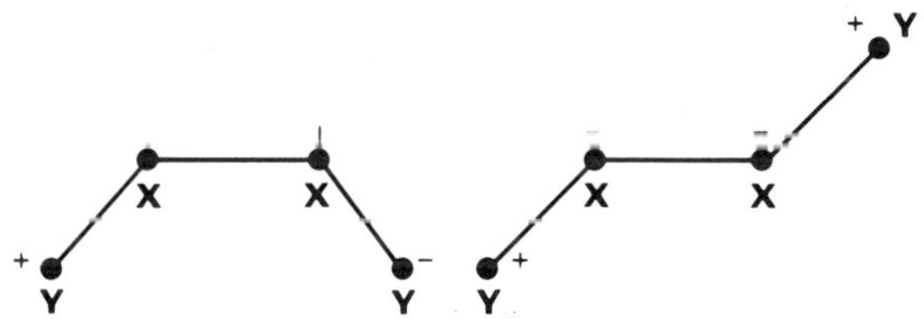

Fig. 10.3 — Illustration of the torsional vibration for YXXY systems (C_{2v} and C_{2h}) ($+ =$ above, $- =$ below the plane of the paper). A torsional force constant f_γ is assigned to every torsional frequency ν_γ.

The torsion appears in the case of four atomic planar YXXY systems (C_{2v} and C_{2h}) as well as in three-dimensional YXXY systems (C_2) in addition to the three valence and two deformation vibrations. It involves four atoms like the out-of-plane vibration (cf. Fig. 10.3).

Summary
The forces acting between chemically bonded atoms are described in the general valence force field for internal co-ordinates by valence force constants (for 2 atoms), deformation force constants (for 3 atoms), out-of-plane force constants (for 4 atoms), and torsional force constants (for 4 atoms). The internal co-ordinates are expressed in terms of bond lengths and angles of the molecule. The influences of forces between non-bonded atoms are included in the coupling force constants. Useful summaries and more extensive treatments can be found in references [65–72].

11

Calculation of the G_r matrix of the kinetic energy for internal co-ordinates

The terms for the kinetic energy of the valence, deformation, out-of-plane, and torsional vibrations are calculated analogously to those of the potential energy. The assignment follows from the corresponding Cartesian co-ordinates $\Delta\mathbf{x}$ for all the N atoms of the molecule, i.e. the vector $\Delta\mathbf{x}$ consisting of the $3N$ Cartesian co-ordinates. The Cartesian co-ordinates are then transformed into internal co-ordinates of the valences, deformations, out-of-plane deformations, and torsions. The vector $\Delta\mathbf{r}$ of the internal co-ordinates for non-linear molecules consists of $3N-6$ elements (6 is the number of translational and rotational degrees of freedom of motion, which drop out as non-vibrations). The relation between the two co-ordinate systems is given by the transformation matrix $\mathbf{B}$:

$$\Delta\mathbf{r} = \mathbf{B} \cdot \Delta\mathbf{x} \tag{11.2}$$

which has $3N-6$ rows and $3N$ columns. This yields the matrix $\mathbf{G}_r$ of the kinetic energy for internal co-ordinates from the matrix $\mathbf{G}_x$ of the kinetic energy for Cartesian co-ordinates $\Delta\mathbf{x}$ according to

$$\mathbf{G}_r = \mathbf{B} \cdot \mathbf{G}_X \cdot \mathbf{B}' \tag{11.2}$$

where

$$\mathbf{G}_X = \mathbf{M}^{-1} = \mathrm{diag}\,(m_i^{-1}) \tag{11.3}$$

and

$$m_i = m_1, m_1, m_1, m_2, m_2, m_2, \ldots, m_N, m_N, m_N$$

(m_1, m_2, ... are the masses of the 1., 2., ... atom of the molecule, and $\mathbf{B}'$ is the transposed $\mathbf{B}$ matrix.)

Whereas the G_x matrix contains only the masses of the atoms, angles, distances and ratios of distances are included in the G_r matrix. This facilitates the interpretations in the chemical context of bond and shape-restoring forces.

The method is used mainly for small molecules [73] since the assembly of the **B** matrix can be quite cumbersome. However, these calculations have shown that, at least in the case of non-cyclic systems,† a building-block approach can be used for the assembly of the G_r matrix. This method has been described by Decius [42]. The methods have been adapted in many spectroscopic treatises because of their practical usefulness. We restrict this discussion to a few frequently occurring cases since the inclusion of the full set of equations would require several pages.

Summary of four equations used in the building-block approach to the assembly of the G_r matrix of the kinetic energy for internal co-ordinates, after Decius, is as follows.

(1) *Valence terms*

    The valence bond for two chemically bonded atoms leads to the sum of the reciprocal masses

$$g_{rr}^2 = \mu_1 + \mu_2 \tag{11.4}$$

    where $\mu = 1/m$. This is the well-known equation of the harmonic oscillator for two atoms (cf. Eq. (7.1.1)) (Fig. 11.1). This way of assembling larger molecules

Fig. 11.1 — The topology of the valence term g_{rr}^2.

    from harmonic oscillators is reflected also in force constant calculations where it helps to estimate the bonding where the coupling is weak or supplies an initial solution as a starting point for the refinement.

(2) *Valence–valence coupling terms*

    The coupling between two neighbouring valences is given by the reciprocal mass of the central atom multiplied by the cosine of the bond angle v (Fig. 11.2)

---

† This approach does not work completely for cyclic systems, but is still useful in many cases.

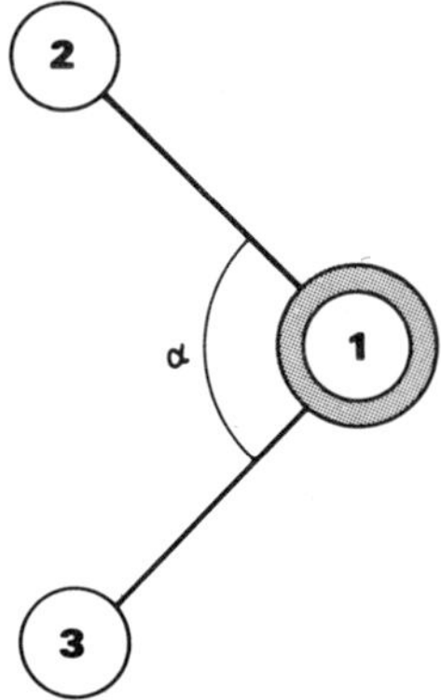

Fig. 11.2 — The topology of the coupling term g_{rr}^1.

$$g_{rr}^1 = \mu_1 \cos \alpha \tag{11.5}$$

By using just these two equations it is already possible to obtain solutions for all molecules which can be described completely by valences and their couplings. These include X_3 (D_{3h}), X_4 (T_d) and X_{12} (I_h).

Example 11.1 X_4 molecules (T_d)
The tetrahedral X_4 (T_d) system (Fig. 11.3) has six (3x4 − 6) vibrational degrees of

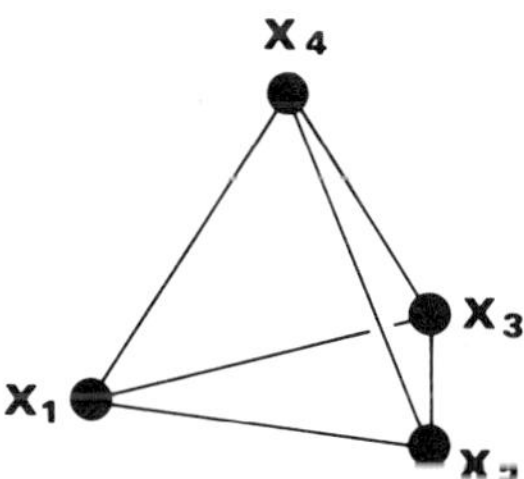

Fig. 11.3 — The geometry of the tetrahedral X_4 (T_d) system.

freedom which can be described by six valence terms along the main diagonal of the G_r matrix corresponding to the six bonds. Thereby all main diagonal elements are defined as $2\mu_X$. The non-zero coupling terms are all $\mu_X/2$ since all bond angles are $60°$.
The resulting G_r matrix can be written as follows:

$$\mathbf{G}_r(\mathbf{X}_4(\mathbf{T}_d)) = \begin{array}{c} \\ \begin{array}{cccccc} 1\text{–}2 & 1\text{–}3 & 2\text{–}3 & 1\text{–}4 & 2\text{–}4 & 3\text{–}4 \end{array} \\ \begin{bmatrix} 2\mu_X & \mu_X/2 & \mu_X/2 & \mu_X/2 & \mu_X/2 & 0 \\ \mu_X/2 & 2\mu_X & \mu_X/2 & \mu_X/2 & 0 & \mu_X/2 \\ \mu_X/2 & \mu_X/2 & 2\mu_X & 0 & \mu_X/2 & \mu_X/2 \\ \mu_X/2 & \mu_X/2 & 0 & 2\mu_X & \mu_X/2 & \mu_X/2 \\ \mu_X/2 & 0 & \mu_X/2 & \mu_X/2 & 2\mu_X & \mu_X/2 \\ 0 & \mu_X/2 & \mu_X/2 & \mu_X/2 & \mu_X/2 & 2\mu_X \end{bmatrix} \begin{array}{c} 1\text{–}2 \\ 1\text{–}3 \\ 2\text{–}3 \\ 1\text{–}4 \\ 2\text{–}4 \\ 3\text{–}4 \end{array} \end{array} \tag{11.6}$$

The P_4 molecule is an example of such a system: $m_P = 30.97$ and therefore $2\mu_X = 0.06458$ and $\mu_X/2 = 0.01614$.

(3) *Deformation terms*

Three reciprocal masses enter the $g^3_{\phi\phi}$ term, since the deformation of a bond angle φ involves three atoms, with the central atom occupying a unique position (Fig. 11.4).

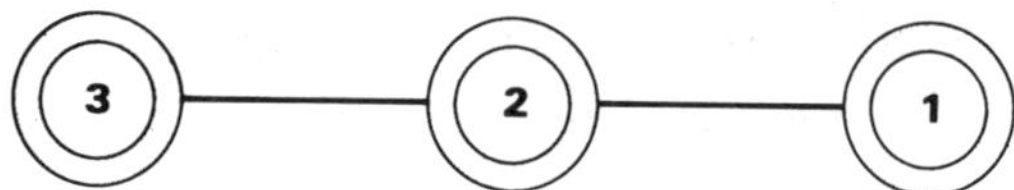

Fig. 11.4 — The topology of the deformation term $g^3_{\phi\phi}$.

This results in the following equation

$$g^3_{\phi\phi} = \rho^2_{12}\mu_1 + \rho^2_{23}\mu_3 + (\rho^2_{12} + \rho^2_{23} - 2\rho_{12}\rho_{23}\cos\varphi)\,\mu_2 \tag{11.7}$$

where $\rho_{12} = r_{12}^{-1}$ and $\rho_{23} = r_{23}^{-1}$ are the reciprocal equilibrium distances and φ is the bond angle.

(4) *Coupling between valence and deformation*

Of the three coupling terms between a valence and a deformation, only the case of two common atoms will be mentioned here (Fig. 11.5).

$$g_{r\vartheta} = -\rho_{23}\mu_2\,\sin\phi \tag{11.8}$$

Example 11.2   The $\mathbf{G}_r$ matrix of the kinetic energy for internal co-ordinates of the system XYZ ($\mathbf{C}_s$)

This general triatomic system (Fig. 11.6) has 3 ($= 3 \times 3 - 6$) internal degrees of freedom. It can be described in terms of the valences X−Y and Y−Z and the deformation XYZ. It is therefore possible to find all elements of the $\mathbf{G}_r$ matrix using Decius' building-block method [42].

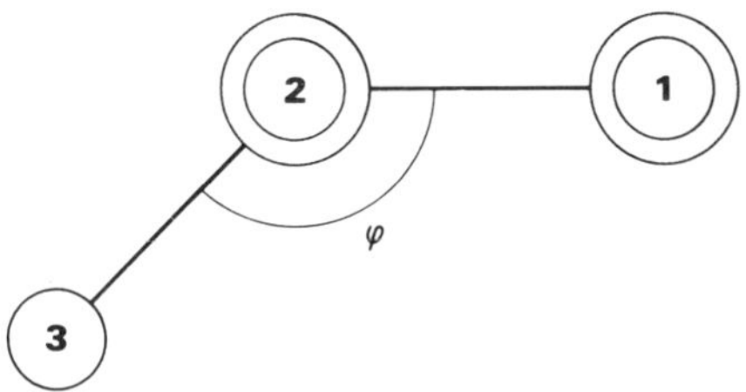

Fig. 11.5 — The topology of the valence–deformation coupling term $g_{r\varphi}^2$.

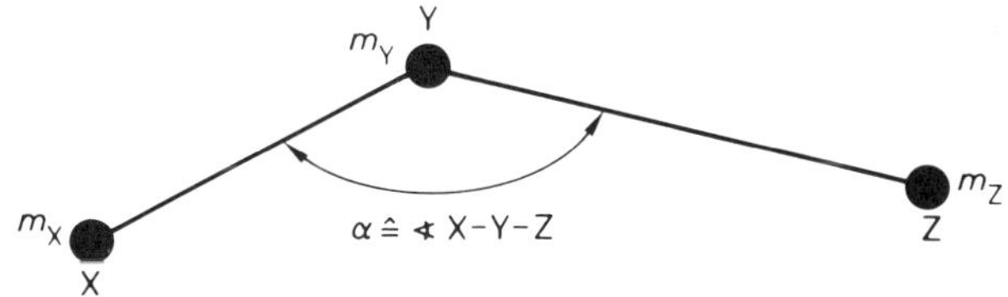

Fig. 11.6 — Geometry and masses of the XYZ (C_S) system.

$$
\mathbf{G}_r(\mathrm{XYZ}(C_S)) = \begin{array}{c} \\ \end{array}
\begin{array}{ccc} X-Y & Y-Z & X-Y \end{array}
\begin{bmatrix} g_{rr}^2 & g_{rr}^1 & g_{r\varphi}^2 \\[4pt] g_{rr}^1 & g_{rr}^2 & g_{r\varphi}^2 \\[4pt] g_{r\varphi}^2 & g_{r\varphi}^2 & g_{\varphi\varphi}^3 \end{bmatrix}
\begin{array}{c} X-Y \\[4pt] Y-Z \\[4pt] X-Y-Z \end{array}
\tag{11.9}
$$

$$
= \begin{bmatrix} \mu_X + \mu_Y & \mu_Y \cos\alpha & -\mu_Y \rho_{XY} \sin\alpha \\[6pt] \mu_Y \cos\alpha & \mu_Y + \mu_Z & -\mu_Y \rho_{YX} \sin\alpha \\[6pt] \mu_Y \rho_{XY} \sin\alpha & -\mu_Y \rho_{YZ} \sin\alpha & g(\mathrm{XYZ}) \end{bmatrix}
\tag{11.9a}
$$

$$
g(\mathrm{XYZ}) = g_{33} = \rho_{YZ}^2 \mu_Z + \rho_{XY}^2 \mu_X + (\rho_{XY}^2 + \rho_{YZ}^2 - 2\rho_{XY}\rho_{YZ}\cos\alpha)\mu_Y \tag{11.9b}
$$

Numerical values for the compounds $^{16}\mathrm{O}^{14}\mathrm{N}^{35}\mathrm{Cl}$, $^{16}\mathrm{O}^{15}\mathrm{N}^{35}\mathrm{Cl}$ and $^{18}\mathrm{O}^{15}\mathrm{N}^{35}\mathrm{Cl}$ can be found in Example 15.1.

The $\mathbf{G}_r$ matrix for the linear system XYZ ($C_{\infty v}$) follows immediately from the above, with $\alpha = 180°$

$$G_r(XYZ(G_{\infty v})) = \begin{bmatrix} \mu_X + \mu_Y & \mu_Y & 0 \\ \mu_Y & \mu_Y + \mu_Z & 0 \\ 0 & 0 & g(XYZ) \end{bmatrix} \qquad (11.10a)$$

$$g(XYZ) \quad = \rho_{YZ}^2 \mu_Z + \rho_{XY}^2 \mu_X + (\rho_{XY} + \rho_{YX})^2 \mu_Y \qquad (11.10b)$$

In this case the molecular geometry leads to a splitting of the G_r (XYZ ($C_{\infty v}$)) matrix of the order $n = 3$ into a matrix of the order $n = 2$ and a matrix of the order $n = 1$.

Numerical values can be found in Examples 13.1, 14.2 (for NNO) and 14.1 (for ClCN).

The G_r matrix for YX_2 ($D_{\infty h}$) follows simply from $Z = X$, which leads to the equations presented in section 7.2 when symmetry co-ordinates are used. By using the masses $m_C = 12.01$ and $m_S = 32.06$, the following parameters are found for the CS_2 molecule:

$$G_{11} = \mu_S = 0.03119 \qquad (11.11)$$

$$G_{22} = \mu_S + 2\mu_C = 0.1977 \qquad (11.12)$$

$$G_{33} = 2(\mu_S + 2\mu_C) = 0.3954 \qquad (11.13)$$

Finally for $Z = X$ and $\alpha \neq 180°$ the matrix obtained is the G_r matrix for the frequently occurring YX_2 (C_{2v}) system, which is the basis for further reduction by using symmetry co-ordinates (cf. Eq. (13.23)). Example 13.2 shows the calculations for the NO_2 molecule. This concludes the description of the G_r matrices of the kinetic energy for internal co-ordinates of triatomic systems, which frequently occur in vibrational spectroscopy.

12

Construction of energy matrices for symmetrical molecules

In Chapter 5 we discussed the effect of molecular symmetry on the separation of vibrations into different species and this is now carried into the matrices G_r of the kinetic energy and F_r of the potential energy for internal co-ordinates. This is done by calculating the energy matrices for internal symmetry co-ordinates, which can be written as a column matrix S.† The symmetry co-ordinates are obtained as a sum of products of the transformation coefficients from the character tables of the point group of the molecule with internal co-ordinates to which the corresponding symmetry operations have been applied. If Δr is the vector of the internal co-ordinates then S follows from

$$S = U \times \Delta r \tag{12.1}$$

where U is the corresponding transformation matrix. Details can be found in the literature.

When the symmetry co-ordinates are known, G_S and F_S the energy matrices for symmetry co-ordinates can be obtained according to (compare Eq. (11.2))

$$F_S = U \cdot F_r \cdot U' \tag{12.2}$$

and

$$G_S = U \cdot G_r \cdot U' \tag{12.3}$$

This reduces the calculation of energy matrices for symmetry co-ordinates to two

---

† A readily comprehensible introduction to this aspect of group theory applied to vibrational spectroscopy can be found in [74] [includes complete calculations for the systems XY_2 (C_{2v}), XY_3 (C_{3v}), XY_4 (T_d), and XY_6 (O_h)].

matrix multiplications, which can, however, be quite cumbersome for larger molecules. For this reason we also point out a 'building-block method', which is quite useful, e.g. for checking the numerical calculations of G_S matrices.

'Building-block method'

In analogy to Decius' method for constructing G_r matrices there are some rules [2] for constructing the energy matrices for symmetry co-ordinates F_S and G_S. We will introduce here only those for the main diagonal elements of the valence terms for the totally symmetric species. In a symmetrical XY_p system all valence force constants $f(XY)$ are equal. The group theoretical reductions derive the symmetry valence element $F(XY)$ from the valence force constant $f(XY)$ for internal co-ordinates and the sum of the coupling constants with the other $p-1$ valences $f(XY/XY)$:

$$F(XY, \text{ totally symmetric species}) = f(XY) + \sum_{i=1}^{p-1} f_i\,(XY/XY) \qquad (12.4)$$

The corresponding **G** matrix elements are found in an analogous fashion:

$$G(XY, \text{ totally symmetric species}) = g(XY) + \sum_{i=1}^{p-1} g_i\,(XY/XY) \qquad (12.5)$$

Example 12.1 $X_4(T_d)$ (cf. Fig. 11.3)
The symmetry force constant for the species A_1 is

$$F_{11}(A_1) = F_{11}(XY) = f(XY) + 4f(XY/XY) + f'(XY/XY) \qquad (12.6)$$

where $f(XY/XY)$ corresponds to the four adjacent valences and $f'(XY/XY)$ to the only opposite valence. The $G(XY)$ element for symmetry co-ordinates can be calculated from the G_r matrix elements given in Eq. (11.6):

$$G_{11}(A_1) = G_{11}(XY) = 2\mu_X + 4\mu_X/2 + 1 \times 0 = 4\mu_X \qquad (12.7)$$

This leads to the equation for the only vibration in the species A_1 with the frequency $v(A_1)$:

$$F_{11}(A_1) \times 4\mu_X = \left[f(XX) + 4f(XX/XX) + f'(XX/XX) \right] \times 4\mu_X = c \times v^2(A_1) \qquad (12.8)$$

The symmetry force constant for the tetrahedral P_4 molecule, with $m_P = 30.97$ and $v(A_1) = 606\ \text{cm}^{-1}\ (Ra)$ [3], is therefore $F_{11}(A_1) = 1.68\ \text{N/cm}$.

Example 12.2   XY$_n$ systems with equivalent, indistinguishable valences
From Eq. (12.4) it follows that for symmetrical systems XY$_n$ with n equivalent, indistinguishable bonds X$-$Y $= r$, the simple and clear relation for the valence force constant of the totally symmetric species is

$$F_r \text{ (totally symmetric species)} = f_r + (n\text{-}1)f_{rr} \qquad (12.9)$$

It holds for the following systems:†
XY$_2$ (**D**$_{\infty h}$); XY$_2$ (**C**$_{2v}$) with $n = 2$,
XY$_3$(**D**$_{3h}$); XY$_3$ (**C**$_{3v}$) with $n = 3$,
XY$_4$ (**T**$_d$) with $n = 4$.

In this way, group theory reduces n equivalent valence force constants to a single constant. Equation (12.9) illustrates in a striking way the theoretical background of group theory in one of its chemical applications. (The same holds for the other symmetry species in the continuation of this method.)

More complex systems can be calculated similarly, rapidly and without computers, as long as only the totally symmetric elements for valences are concerned.

Based on the procedures for the symmetry co-ordinate system and the corresponding transformations, all other elements of the energy matrices $\mathbf{F}_S$ and $\mathbf{G}_S$, also those for non-symmetric species, can be calculated. More extensive discussions can be found in the literature [77, 78].

---

† Eq. (12.9) can also be applied to the system X$_3$ (**D**$_{3h}$) $n = 3$.

IV.C
Calculation of the force constant matrix for $n = 2$

The lowest order of the vibrational equation for unsymmetrical systems is $n = 3$ for 3-atom molecules, $n = 6$ for 4-atom molecules and $n = 9$ for 5-atom molecules. These split only for symmetrical systems (and larger polyatomic molecules) into eigenvalue problems of the orders $n = 1$ and 2.

The calculation of a force constant matrix of the order $n = 2$ illustrates clearly and on a small scale the problems involved in the calculation of such matrices for an arbitrary system. Here we will deal only with two important cases:

The vibrational frequencies of an isotopic system are known. This leads to the complete, exact calculation of the elements f_{11}, f_{22} and f_{12} of the force constant matrix (Chapter 13)

$$\mathbf{F} = \begin{bmatrix} f_{11} & f_{12} \\ f_{12} & f_{22} \end{bmatrix}$$

Only the two vibrational frequencies v_1 and v_2 of the system are known, which is most frequently the case. In this case extremal solutions can be calculated, and these can provide useful results for numerous chemical applications (Chapter 14).

13

Exact calculation of force constants for known isotope frequencies

The computation of the three unknown force constants f_{11}, f_{22} and f_{12} of the force constant matrix of the order $n = 2$ requires in addition to the two equations from the secular equation a further equation which is often obtained from the vibrational frequencies of an isotopic molecule.

$$g_{11}f_{11} + g_{22}f_{22} + 2g_{12}f_{12} = \lambda_1 + \lambda_2 \tag{13.1}$$

$$(g_{11}g_{22} - g_{12}^2)\,(f_{11}f_{22} - f_{12}^2) = \lambda_1 \times \lambda_2 \tag{13.2}$$

$$g'_{11}f_{11} + g'_{22}f_{22} + 2g'_{12}f_{12} = \lambda'_1 + \lambda'_2 \tag{13.3}$$

(The isotopic data are indicated by a prime $'$.)

The solution of the two linear and the one quadratic equation leads necessarily to a quadratic equation for one of the three unknown force constants, so that the two force constant matrices $\mathbf{F}^+$ and $\mathbf{F}^-$ are mathematically possible solutions. The explicit calculation is shown in the following set of equations [79].

$$\mathbf{F}^+ = \begin{bmatrix} f_{11}^+ & f_{12}^+ \\ f_{12}^+ & f_{22}^+ \end{bmatrix} \tag{13.4}$$

$$\mathbf{F}^- = \begin{bmatrix} f_{11}^- & f_{12}^- \\ f_{12}^- & f_{22}^- \end{bmatrix} \tag{13.5}$$

$$f_{11}^{\pm} = \frac{-B \pm \sqrt{B^2 + 4AC}}{2A} \tag{13.6}$$

$$f_{22}^{\pm} = \frac{K_6 - K_4 f_{11}^{\pm}}{K_5} \tag{13.7}$$

$$f_{12}^{\pm} = \frac{K_2 - g_{11} f_{11}^{\pm} - g_{22} f_{22}^{\pm}}{2g_{12}} \tag{13.8}$$

$$K_1 = \lambda_1 \lambda_2 \det^{-1} \mathbf{G} \quad \text{where} \quad \det \mathbf{G} = g_{11} g_{22} - g_{12}^2 \tag{13.9}$$

$$K_2 = \lambda_1 + \lambda_2 \tag{13.10}$$

$$K_3 = \lambda_1' + \lambda_2' \tag{13.11}$$

$$K_4 = g_{11} g_{12}' - g_{11}' g_{12} \tag{13.12}$$

$$K_5 = g_{22} g_{12}' - g_{22}' g_{12} \tag{13.13}$$

$$K_6 = g_{12}' K_2 - g_{12} K_3 \tag{13.14}$$

$$A = -g_{11}^2 + K_4 K_5^{-1}\left(2(\det \mathbf{G} - \mathbf{g}_{12}^2) - g_{22}^2 K_4 K_5^{-1}\right) \tag{13.15}$$

$$B = K_6 K_5^{-1}\left(2(g_{12}^2 - \det \mathbf{G}) + 2g_{22}^2 K_4 K_5^{-1}\right) + 2K_2(g_{11} - g_{22} K_4 K_5^{-1}) \tag{13.16}$$

$$C = -K_2^2 - 4K_1 g_{12}^2 + K_5 K_6^{-1} g_{22}(2K_2 - g_{22} K_6 K_5^{-1}) \tag{13.17}$$

with the condition that $K_5 \neq 0$ $\tag{13.18}$

When $A = 0$ then $f_{11} = -CB^{-1}$ $\tag{13.19}$

$$\lambda_i = 5.89146 \times 10^{-7} \times v_i^2 \quad \text{for} \quad i = 1,2 \tag{13.20}$$

The selection of the physicochemically meaningful solution requires additional knowledge. Sometimes it is sufficient to calculate the force constants of similar diatomic molecules or to compare the results with those obtained for the uncoupled approximation.

$$f_{ii} = \lambda_i / g_{ii} \quad \text{for } i = 1, 2 \tag{13.21}$$

Example 13.1   Force constants of NNO ($\mathbf{C}_{\infty v}$)
Calculation of the valence force constants f_{NN}, f_{NO} and the deformation force constant $f_{NN/NO}$ of the linear NNO molecule by using the $\mathbf{G}$ matrices given in Eq. (11.10) ($n = 2$) and an additional set of isotope frequencies.
Infrared frequencies for the molecules $^{14}N^{14}N^{16}O$ and $^{15}N^{14}N^{16}O$ [80] are

$$N(^{14}N^{14}N^{16}O) = \begin{bmatrix} 2282 & \\ & 1299 \end{bmatrix}$$

$$N(^{15}N^{14}N^{16}O) = \begin{bmatrix} 2259 & \\ & 1282 \end{bmatrix} \text{cm}^{-1}$$

The following G_S matrices† are obtained:

$$G_S(^{14}N^{14}N^{16}O) = \begin{bmatrix} 0.1428 & -0.07141 \\ -0.07141 & 0.1339 \end{bmatrix}$$

$$G_S(^{15}N^{14}N^{16}O) = \begin{bmatrix} 0.1381 & -0.07141 \\ -0.07141 & 0.1339 \end{bmatrix}$$

The calculation leads to the two solutions‡

$$F^- = \begin{bmatrix} 18.59 & 18.64 \\ 18.64 & 30.39 \end{bmatrix}$$

and

$$F^+ = \begin{bmatrix} 18.59 & 1.19 \\ 1.19 & 11.78 \end{bmatrix} \text{N/cm}$$

The solution F^- can be eliminated right away because the coupling constant $f_{NN/NO} = 18.64$ N/cm and the valence force constant $f_{NO} = 30.39$ N/cm are much too large. This establishes

$$F_S(NNO) = \begin{bmatrix} f_{NN} & f_{NN/NO} \\ f_{NN/NO} & f_{NO} \end{bmatrix} = \begin{bmatrix} 18.59 & 1.19 \\ 1.19 & 11.78 \end{bmatrix} \text{N/cm} \qquad (13.22)$$

as the chemically meaningful force constant matrix. With the isotope reduction method, which has been described in Chapter 8, the same data give $f_{NN} = 18.49$ N/cm, and with additional data [80] $f_{NO} = 11.55$ N/cm (cf. Table A.1).

† For symmetry reasons, $G_r = G_S$, which is especially important for the F_S matrix where as a consequence all valence–deformation coupling elements are zero.

To obtain $K_S \neq 0$ it is advisable to exchange g_{11} and g_{22} in both G_S matrices.

The NNO molecule shows strong vibrational coupling, $f_{NO/NO} = 1.19$ N/cm, which is caused by the fact that all three masses are almost equal. This is also very obvious in a comparison with the uncoupled values for the valence force constants $f_{NN,unc} = 21.48$ and $f_{NO,unc} = 7.42$ N/cm.

An interpretation in terms of bonding and mesomeric structures can be found in Chapter 20.

Example 13.2   Force constants of ONO (C_{2v})

Calculation of the valence, deformation, and all coupling force constants, in N/cm, of the bent ONO molecule (C_{2v}). The G_r matrix element for the deformation $g(ONO)$ has according to Eq. (11.10a) the dimensions of inverse mass times inverse squared length. Since the valence deformation coupling terms have the dimensions of inverse mass times inverse length, the deformation force constant must have the units N·cm and the valence deformation coupling constant, N. To give all matrix elements of the G_S and F_S matrices the same units they can be written as follows [81]:

$$G_S(XY_2) = \begin{bmatrix} \mu_X(1+\cos\alpha)+\mu_Y & -\sqrt{2}\,\mu_X\sin\alpha & 0 \\ -\sqrt{2}\,\mu_X\sin\alpha & 2(\mu_Y+\mu_X(1-\cos\alpha)) & 0 \\ 0 & 0 & \mu_Y+\mu_X(1-\cos\alpha) \end{bmatrix} \text{amu}^{-1}$$

(13.23)

$$F_S(XY_2) = \begin{bmatrix} f_{XY}+f_{XY/XY} & \sqrt{2}\,f_{XY/\alpha} & 0 \\ \sqrt{2}\,f_{XY/\alpha} & f_\alpha & 0 \\ 0 & 0 & f_{XY}-f_{XY/XY} \end{bmatrix} \text{N/cm}$$

(13.24)

The following data are given for the isotopic molecules $^{14}NO_2$ and $^{15}NO_2$:

$m_{14_N} = 14.003$, $m_{15_N} = 15.000$, $m_0 = 15.999$; $\alpha = 134.3°$ [82]; $r_{NO} = 0.1197$ nm (however, this is not required for force constant calculations in N/cm); The infrared frequencies in cm^{-1} [83,84] are

	Species A_1		Species B_2
NO$_2$	ν_1	ν_2	ν_3
$^{14}NO_2$	1320	750	1618
$^{15}NO_2$	1307	740	1580

The elements of the G_S matrices are

$^{14}NO_2$: $G_{11} = 0.08404$, $G_{12} = -0.07228$,
$\quad\quad\quad G_{22} = 0.3676$, $G_{33} = 0.1838$.
$^{15}NO_2$: $G_{11} = 0.08261$, $G_{12} = -0.06748$,
$\quad\quad\quad G_{22} = 0.3515$, $G_{33} = 0.1757$.

The symmetry force constant

$$F_{33} = f_{XY} - f_{XY/XY} = f_{NO} - f_{NO/NO}$$

of the species B_2 is calculated according to the two-mass model (cf. section 7.1). The three symmetry force constants

$$F_{11} = f_{NO} + f_{NO/NO}, \qquad F_{22} = f_\alpha$$

and

$$F_{12} = \sqrt{2}\, f_{NO/\alpha}$$

of the species A_1 are obtained according to Eqs. (13.1) to (13.21). The following chemically meaningful solution is obtained.

$$\mathbf{F}_S = \begin{bmatrix} 12.58 & 0.69 & 0 \\ 0.69 & 1.09 & 0 \\ 0 & 0 & 8.39 \end{bmatrix} \text{N/cm}$$

This leads to the following force constants for internal co-ordinates:

$$f_{NO} = 10.48; \quad f_{NO/NO} = 2.10; \quad f_\alpha = 1.09; \quad f_{NO/\alpha} = 0.49 \text{ N/cm}.$$

14

Extremal force constants for the case of only two known frequencies — method of the next solution

In a large number of cases only two frequencies are known and isotopic systems are not available. If the couplings are small, a reasonable answer is found from the two diagonal force constants f_{11} and f_{22}, with f_{12} set equal to zero. This is not possible if the coupling is stronger. The numerical solutions will then be strongly divergent. Complex solutions may even be obtained.

One way out of this dilemma is the use of extremal force constants, which can be obtained in various ways [85]. We will introduce here only the method of the next solution [86], which works as follows.

Force constant matrices $\mathbf{F}_S$ are selected with a minimizing condition from the infinite range of solutions for $\mathbf{F}_S$, according to the secular equation $\det(\mathbf{G}\cdot\mathbf{F}_S - \lambda\mathbf{E}) = 0$, by adding an approximate solution $\mathbf{F}_{ap}$. For physicochemical applications it is best to use the analytical solution that can be obtained by assuming the normal vibrations to be completely uncoupled†, if the couplings in $\mathbf{F}$ are small. For larger couplings the exact force constants of smaller molecules can be used as a starting (approximate) solution.

Conversion:

$$\lambda_i = 5.89146 \times 10^{-7} \times v_i \text{ for } i = 1, 2 \tag{14.1}$$

Validity range:

$$f_{12,\text{ext}}^{\pm} = f_{12}^{\pm} = \frac{c_1 g_{12} \pm (\lambda_1 - \lambda_2)\sqrt{g_{11}g_{22}}}{2 \det \mathbf{G}} \tag{14.2}$$

† The uncoupled solution contains force constants of a system of uncoupled harmonic oscillators.

with

$$c_1 = -(\lambda_1 + \lambda_2) \tag{14.3}$$

$$c_0 = \lambda_1 \times \lambda_2 \tag{14.4}$$

$$\det \mathbf{G} = g_{11}g_{22} - g_{12}^2 \tag{14.5}$$

Solutions from the infinite range of possible solutions:

$$\mathbf{F}_S = \begin{bmatrix} f_{11}^+ & f_{12}^\pm \\ f_{12}^\pm & f_{22}^- \end{bmatrix} \tag{14.6}$$

$$\mathbf{F}_S = \begin{bmatrix} f_{11}^- & f_{12}^\pm \\ f_{12}^\pm & f_{22}^+ \end{bmatrix} \tag{14.7}$$

with

$$f_{11}^\pm = g_{11}^{-1} \times \mathbf{K}^\pm \tag{14.8}$$

$$f_{22}^\pm = g_{22}^{-1} \times K^\pm \tag{14.9}$$

$$K^\pm = 0.5(-(c_1 + 2g_{12}f_{12}) \pm \sqrt{(c_1 + 2g_{12}f_{12})^2 - 4g_{11}g_{22}(f_{12}^2 + c_0 \det^{-1} \mathbf{G}))} \tag{14.10}$$

$$f_{12} \quad \text{corresponds to } f_{12}^\mp \text{ or } f_{12}^\pm$$

Approximate solutions as starting solution
Uncoupled vibrations

$$\mathbf{F}_{ap} = \mathbf{F}_{un} = \begin{bmatrix} g_{11}^{-1}\lambda_1 & \\ & g_{22}^{-1}\lambda_2 \end{bmatrix} \tag{14.11}$$

or physicochemical approximation

$$\mathbf{F}_{ap} = \mathbf{F}_{phys\text{-}chem} = \begin{bmatrix} f_{ap,11} & f_{ap,12} \\ f_{ap,12} & f_{ap,22} \end{bmatrix} = \begin{bmatrix} f_{11,n} & f_{12,n} \\ f_{12,n} & f_{22,n} \end{bmatrix} \tag{14.12}$$

Minimizing condition

$$\frac{dQ}{df_{12}} = 0 \tag{14.13}$$

with

$$Q = (f_{11,n} - f_{11})^2 + (f_{22,n} - f_{22})^2 + (f_{12,n} - f_{12})^2 \tag{14.14}$$

The minimum is found by iteration.

Example 14.1   Extremal force constants of ClCN

Calculation of the valence force constants f_{CIC} and f_{CN} and the coupling constant $f_{CIC/CN}$ by using the uncoupled case as a starting point. The infrared frequencies are taken from the literature [87].

$$\mathbf{N} = \begin{bmatrix} \nu_{CIC} & \\ & \nu_{CN} \end{bmatrix} = \begin{bmatrix} 714 & \\ & 2219 \end{bmatrix} \text{cm}^{-1} \tag{14.15}$$

The $\mathbf{G}_S$ matrix of the linear system is, according to Eq. (11.10),

$$\mathbf{G}_S = \begin{bmatrix} 0.1115 & -0.08326 \\ -0.08326 & 0.1547 \end{bmatrix} \tag{14.16}$$

The solution range for the coupling force constant $f_{CIC/CN}$ lies between -3.63 and 29.47 N/cm. From the solution for the uncoupled case one obtains

$$\mathbf{F}_{de} = \begin{bmatrix} 2.69 & \\ & 18.75 \end{bmatrix} \tag{14.17}$$

as a chemically plausible starting solution for the iterations (method of the next solution). Thus one obtains the extremal force constants

$$\mathbf{F}_{S,ex} = \begin{bmatrix} f_{CIC} & f_{CIC/CN} \\ f_{CIC/CN} & f_{CIC} \end{bmatrix} = \begin{bmatrix} 4.65 & 0.92 \\ 0.92 & 18.33 \end{bmatrix} \text{N/cm} \tag{14.18}$$

These lie within or close to the range of force constants that have been obtained with complete sets of data [14]:

$$f_{CIC} \text{ from 4.74 to 5.26,} \quad f_{CN} \text{ from 16.55 to 18.55} \quad \text{and}$$
$$f_{CIC/CN} \text{ from 0.23 to 1.34 N/cm}$$

These approximate force constants should be sufficiently accurate for a number of applications, such as the assignment of frequencies, provided that the vibrations are not coupled too strongly.

Example 14.2   Extremal force constants of NNO

This molecule with strongly coupled vibrations which has been already considered in section 13.1 will now be used to illustrate the limitations of using the approximation of uncoupled vibrations as a starting point for the iterations. The data are taken from $\mathbf{N}(^{14}\mathrm{N}^{14}\mathrm{N}^{16}\mathrm{O})$ and $\mathbf{G}_S(^{14}\mathrm{N}^{14}\mathrm{N}^{16}\mathrm{O})$. The result is

$$\mathbf{F}_{S,\mathrm{ex}} = \begin{bmatrix} f_{\mathrm{NN}} & f_{\mathrm{NN/NO}} \\ f_{\mathrm{NN/NO}} & f_{\mathrm{NO}} \end{bmatrix} = \begin{bmatrix} 20.93 & 2.51 \\ 2.51 & 10.69 \end{bmatrix} \mathrm{N/cm} \qquad (14.19)$$

In this case the valence force constants differ by about 10% from those obtained with complete sets of data.

IV.D
Calculation of the force constants of arbitrary molecules

The calculation of force constants for triatomic molecules XYZ with C_s symmetry already involves a problem of the order $n = 3$ with six unknown force constants in

$$\mathbf{F} = \begin{bmatrix} f_{11} & f_{12} & f_{13} \\ f_{21} & f_{22} & f_{23} \\ f_{31} & f_{32} & f_{33} \end{bmatrix}$$

For N-atom, non-linear systems the order increases rapidly according to $n = 3N - 6$. These force constant calculations can be solved in most cases only by iterative methods, i.e. no longer analytically (compare Chapter 8), even if a sufficient amount of isotope data is available.

First we will treat the calculations with isotope data sets, although these are rarely complete.

Then we will discuss the calculation of extremal force constants, which is often the only method to calculate the n vibrational frequencies.

15

Iterative calculations when sufficient isotope frequencies are known

15.1 FORCE CONSTANT MATRICES OF ARBITRARY ORDER

Exactly $n(n+1)/2$ algebraically independent equations from the secular equation for the molecule and its isotopic compositions are required for the complete iterative calculation of the $n(n+1)/2$ force constants of the force constant matrix

$$
\mathbf{F} = \begin{bmatrix} f_{11} & f_{12} & \cdots & f_{1n} \\ f_{21} & f_{22} & \cdots & f_{2n} \\ \cdots & \cdots & \cdots & \cdots \\ f_{n1} & f_{n2} & \cdots & f_{nn} \end{bmatrix}
\tag{15.1.1}
$$

The algebraically dependent equations have to be excluded, that is $\det \mathbf{F} = \det \Lambda / \det \mathbf{G}$ can be used only once since $\mathbf{F}$ is always the same (cf. Eq. (9.3.6)).

Johansen (among others) has developed several methods for accomplishing this task [89,90].

15.2 MATRICES OF ORDER $n = 3$

Complete force constant calculations exist, with few exceptions, only for the order $n = 3$, e.g. for the systems XYZ (C_s), XY_2Z (C_{2v}) and $ZXY_3(C_{3v})$ [91]. For this task, three isotopic systems with three vibrational frequencies each have to be known. This leads to a system of three linear and three quadratic equations, and one cubic equation. Since seven equations are normally available for the calculation of six unknown force constants, one of them (e.g. the cubic equation) can be used for checking purposes.

Explicitly, the secular equation yields the following:

$$g^i_{11}f_{11} + g^i_{22}f_{22} + g^i_{33}f_{33} + 2(g^i_{12}f_{12} + g^i_{23}f_{23} + g^i_{13}f_{13}) = \lambda^i_1 + \lambda^i_2 + \lambda^i_3 \quad (15.2.1)$$

$$(g^i_{11}g^i_{22} - g^{i2}_{12})(f_{11}f_{22} - f^2_{12}) + (g^i_{11}g^i_{33} - g^{i2}_{13})(f_{11}f_{33} - f^2_{13}) +$$
$$(g^i_{22}g^i_{33}g^{i2}_{23})(f_{22}f_{33} - f^2_{23}) + 2(g^i_{11}g^i_{23} - g^i_{12}g^i_{13})(f_{11}f_{23} - f_{12}f_{13}) +$$
$$2(g^i_{22}g^i_{13} - g^i_{12}g^i_{23})(f_{22}f_{13} - f_{12}f_{23}) + 2(g^i_{33}g^i_{12} - g^i_{13}g^i_{23})(f_{33}f_{12} - f_{13}f_{23})$$
$$= \lambda^i_1\lambda^i_2 + \lambda^i_1\lambda^i_3 + \lambda^i_2\lambda^i_3 \quad (15.2.2)$$

where i is used to indicate the different isotopic systems, i.e. $i = $ (blank) original system; $i = {}'$ first isotopic system; $i = {}''$ second isotopic system.

This system of equations leads to eight solutions for one unknown force constant, from which the chemically meaningful force constant has to be chosen by using further criteria.

The cubic equation

$$f_{11}f_{22}f_{33} - f^2_{13}f_{22} - f^2_{23}f_{11} - f^2_{12}f_{33} + 2f_{12}f_{13}f_{23} = \frac{\lambda_1\lambda_2\lambda_3}{\det \mathbf{G}}$$

$$= \frac{\lambda_1\lambda_2\lambda_3}{g_{11}g_{22}g_{33} - g^2_{13}g_{22} - g^2_{23}g_{11} - g^2_{12}g_{33} + 2g_{12}g_{13}g_{23}} \quad (15.2.3)$$

is available for special purposes, such as the calculation of geometrical parameters.

Approximate calculation for the system XYZ (C_s)

For triatomic XYZ (C_s) systems a strong valence coupling constant $f_{XY/YZ} = f_{12}$ is expected, whereas the coupling constants $f_{XY/\alpha} = f_{13}$ and $f_{YZ/\alpha} = f_{23}$ are expected to be much smaller. Therefore we set

$$f_{XY/\alpha} = f_{YZ/\alpha} = 0 \quad (15.2.4)$$

so that the four force constants

$$f_{XY} = f_{11}, \quad f_{YZ} = f_{22}, \quad f_\alpha = f_{33}, \quad \text{and} \quad f_{XY/YZ} = f_{12} \quad (15.2.5)$$

remain to be calculated. We furthermore assume that in addition to the data for XYZ those for the isotopic system X'YZ are also known. One can then use the following approximation, consisting of the three linear equations plus the cubic equation which can be solved by pocket calculators or microcomputers:

Step 1:
$$f_{33} = \frac{\lambda_3}{g_{33}} \text{ as starting value} \quad (15.2.6)$$

Step 2:   The three linear equations are rewritten to obtain the force constants f_{11}, f_{22} and f_{12}:

$$f_{11} = \frac{(Sp\,(\Lambda - \Lambda'))}{(\mu_X - \mu_{X'})} - f_{33}/\upsilon \tag{15.2.7}$$

$$f_{22} = \frac{R'g''_{12} - R''g'_{12}}{g'_{22}g''_{12} - g''_{22}g'_{12}} \tag{15.2.8}$$

with

$$\upsilon = \frac{r_{XY}}{r_{YZ}} \tag{15.2.8a}$$

$$R' = \lambda'_1 + \lambda'_2 + \lambda'_3 - g'_{11}f_{11} - g'_{33}f_{33} \tag{15.2.9}$$

$$R'' = \lambda''_1 + \lambda''_2 + \lambda''_3 - g''_{11}f_{11} - g''_{33}f_{33} \tag{15.2.10}$$

and

$$f_{12} = \frac{Sp\Lambda - g_{11}f_{11} - g_{22}f_{22} - g_{33}f_{33}}{2g_{12}} \tag{15.2.11}$$

Step 3:   The four force constants obtained in this way are then evaluated by using the cubic equation

$$(f_{11}f_{22} - f_{12}^2)f_{33} - \det \Lambda \times \det^{-1} G = 0 \tag{15.2.12}$$

If the left side is less than zero, then the force constant f_{33} in step 1 has to be increased. If the left side is greater than zero, then a smaller value has to be chosen for f_{33}. In this way an approximate force constant matrix **F** with four force constants can be obtained in a few iterations.

Example 15.1   Force constants of NOCl
The order of the energy matrices of the triatomic, bent NOCl molecule is 3 ($= 3 \times 3 - 6$). Therefore the vibrational frequencies of three isotopic NOCl molecules are required for the force constant calculation. The infrared frequencies [cm^{-1}] for the three systems $^{16}O^{14}N^{35}Cl$, $^{16}O^{15}N^{35}Cl$ and $^{18}O^{15}N^{35}Cl$ are taken from the literature [50,92].

$^{16}O^{14}N^{35}Cl$	$^{16}O^{15}N^{35}Cl$	$^{18}O^{15}N^{35}Cl$	Assignment
1836	1804	1753	ON
603	589	581	NCl
336	334	327	$ONCl = \alpha$

The **G** matrix has already been given in Chapter 11 (Eq. (11.9a)). The deformation terms have to be rewritten to obtain uniform units:

$$g_{13} = - v\mu_Y \sin \alpha \tag{15.2.13}$$

$$g_{23} = - v^{-1}\mu_Y \sin \alpha \tag{15.2.14}$$

$$v = \frac{r_{XY}}{r_{YZ}} \tag{15.2.15}$$

$$g_{33} = v^{-1}\mu_X + v\mu_Z + \mu_Y(v^{-1} + v - 2\cos \alpha) \tag{15.2.16}$$

With use of

$$v = \frac{0.114}{0.198} = 0.575_8; \qquad \alpha = 113° \text{ [6]}$$

$$m_{16_O} = 15.995; \qquad m_{18_O} = 17.999; \qquad m_{14_N} = 14.003; \qquad m_{15_N} = 15.000;$$

$$m_{35_{Cl}} = 34.969$$

we obtain:†

$$\mathbf{G}(^{16}O^{14}N^{35}Cl): \quad \begin{aligned} g_{11} &= 0.1339; & g_{12} &= -0.02790 \\ g_{13} &= -0.03785; & g_{22} &= 0.1000 \\ g_{23} &= -0.1142; & g_{23} &= 0.3460 \end{aligned}$$

$$\mathbf{G}(^{16}O^{15}N^{35}Cl): \quad \begin{aligned} g_{11} &= 0.1292; & g_{12} &= -0.02605 \\ g_{13} &= -0.03533; & g_{22} &= 0.09526; \\ g_{23} &= -0.1066; & g_{33} &= 0.3313 \end{aligned}$$

† It is advisable for practical purposes to include even more decimal places for the **G** matrix elements, because the data sets are so similar in the isotope calculations which leads to a loss of decimal places (compare accuracy of the distances and angles).

$$\mathbf{G}(^{18}\mathrm{N}^{15}\mathrm{N}^{35}\mathrm{Cl}):$$

$$g_{11} = 0.1222; \qquad g_{12} = -0.02605$$
$$g_{13} = -0.03533; \qquad g_{22} = 0.09526$$
$$g_{23} = -0.1066; \qquad g_{33} = 0.3192$$

The force constant calculations are now done by five different methods:

(1) *Uncoupled vibrations approximation*

$$f_{\mathrm{ON}} = 14.83; \qquad f_{\mathrm{NCl}} = 2.14; \qquad f_{\alpha} = 0.19\,\mathrm{N/cm}$$

(cf. Eq. (7.1.1) and section 7.4)

(2) *Analytical calculation of the upper limit for f_{ON} according to the isotope reduction method* (Chapter 8)

From the difference between the two linear equations for $^{16}\mathrm{O}^{15}\mathrm{N}^{35}\mathrm{Cl}$ and $^{18}\mathrm{O}^{15}\mathrm{N}^{35}\mathrm{Cl}$ it follows that, with $f_{\alpha} = 0$ as minimum value, the exact upper limit for the valence force constant $f_{\mathrm{NO,\,upper\,limit}}$ is $16.3\,\mathrm{N/cm}$.

(3) *Solving the three linear equations*

Solving the three linear equations (Eq. (15.2.1) by setting g_{12}, g_{13}, g_{23} and f_{12}, f_{13}, f_{23} equal to zero gives

$$f_{\mathrm{ON}} = 18; \qquad f_{\mathrm{NCl}} = 1.8 \qquad \text{and} \qquad f_{\alpha} = -0.8\,\mathrm{N/cm}$$

The deformation force constant is already negative, and f_{ON} is far above the upper limit. The wrong sign of f_{α} indicates strong vibrational coupling (cf. section 9.3 and Eq. (9.3.1)).

(4) *Approximate calculations*

The approximations† for the determination of the four force constants f_{ON}, $f_{\mathrm{NCl}}, f_{\alpha}$, and $f_{\mathrm{ON/NCl}}$ according to Eqs. (15.2.4) to (15.2.12) lead to‡

---

† A further approximation can be made by splitting the energy matrices of the order $n = 3$ into several of the order $n = 2$ by using Eqs. (13.1) to (13.21). This gives

$$\mathbf{G}_r = \begin{bmatrix} g_{\mathrm{ON}} & g_{\mathrm{ON/NCl}} \\ g_{\mathrm{ON/NCl}} & g_{\mathrm{NCl}} \end{bmatrix} \qquad\qquad (15.2.17)$$

and the force constants

$$f_{\mathrm{ON}} = 15.10; \qquad f_{\mathrm{NCl}} = 2.29 \quad \text{and} \quad f_{\mathrm{ON/NCl}} = 0.91\,\mathrm{N/cm}$$

The other two resulting matrices give the values

$$f_{\mathrm{NCl}} = 2.82; \qquad f_{\alpha} = 0.32 \qquad \text{and} \qquad f_{\mathrm{NCl/\alpha}} = 0.49\,\mathrm{N/cm}$$

and

$$f_{\mathrm{ON}} = 15.30; \qquad f_{\alpha} = 0.33 \qquad \text{and} \qquad f_{\mathrm{ON/\alpha}} = 1.46\,\mathrm{N/cm}.$$

‡ The calculations were done with additional decimals in the **G** matrix terms and one additional decimal in the frequency values because of the very similar **G** matrices and frequencies [92].

$$f_{ON} = 15.53; \quad f_{NCl} = 1.75; \quad f_\alpha = 0.44 \quad \text{and} \quad f_{ON/NCl} = 2.56 \, \text{N/cm}$$

(5) *Complete calculation*

The complete calculation is done according to Eqs. (15.2.1) to (15.2.3). The result is [92]

$$f_{ON} = 15.26; \quad f_{NCl} = 1.27; \quad f_\alpha = 0.58; \quad f_{ON/NCl} = 1.53$$

$$f_{ON/\alpha} = 0.09 \quad \text{and} \quad f_{NCl/\alpha} = 0.06 \, \text{N/cm}$$

16

Calculations of extremal force constants of molecular systems with *n* vibrational frequencies — the stepwise coupling method

There are about 20 systems of the order $n = 3$ which can be calculated exactly. Only n frequencies are known for most other molecular systems which is insufficient to calculate the $n(n + 1)/2$ force constants since $n(n - 1)/2$ are missing. As in the case of the order $n = 2$ it is possible to calculate extremal force constants by introducing additional constraints, for which we use chemically appropriate initial assumptions (cf. Chapter 14).

The method of the next solution and the closely related stepwise coupling method have been used to calculate thousands of molecular systems. The starting point is the approximation of a system of uncoupled harmonic oscillators $f_{ii} = \lambda_i/g_{ii}$ with $i = 1$, $2, \ldots, n$. The approximate solutions for the algebraic equations are obtained by increasing in a stepwise manner the coupling in the **G** matrix with the Newton method. This reduces the range of solutions to one guided solution. The sets of equations for this method can be found in the literature [94–97].

Calculations of extremal force constants for many symmetries, according to the stepwise coupling method, can be found, for example, in [98].

Example 16.1   Extremal force constants for NOCl
The extremal force constants for the triatomic molecule $O^{14}NCl$ (C_s) will be calculated by the stepwise coupling method starting from the approximation of uncoupled vibrations.

The **G** matrix is calculated according to Eq. (11.9a) from $r_{ON} = 0.14$ and $r_{NCl} = 0.195$ nm and $\alpha = 116°$.

The extremal force constant matrix $\mathbf{F}_r$ is as given by [96]

$$\mathbf{F}_r = \begin{bmatrix} f_{ON} & f_{ON/NCl} & f_{ON/\alpha} \\ f_{ON/NCl} & f_{NCl} & f_{NCl/\alpha} \\ f_{ON/\alpha} & f_{NCl/\alpha} & f_\alpha \end{bmatrix} = \begin{bmatrix} 14.13 & 0.07 & 0.06 \\ 0.07 & 2.25 & 0.13 \\ 0.06 & 0.13 & 0.30 \end{bmatrix} \text{N/cm} \quad (16.1)$$

A comparison with the results obtained from the complete data set in Example 15.1 shows a significant difference for the valence force constant f_{NCl} and the deformation force constant f_α. Even the largest valence force constant f_{ON} shows a deviation of about 8% from the exact value 15.26 N/cm.

IV.E
Applications and interpretations

The different kinds of frequency calculations and assignments play an important role in the area of applications (Chapter 17).

Many trends can be observed by comparing the results for series of homologous compounds; some of these are listed here (Chapter 18).

To obtain a first impression of the bonding in a new compound the estimation of bond length by interpolation according to Badger is frequently used (Chapter 19).

Finally we mention Siebert's method for estimating the bond order which has been used successfully in many structural problems to interpret results by this simple bonding model (Chapter 20).

17

Calculation of vibrational frequencies

The computation of frequencies of isotopic systems and the estimation of frequencies to support assignments are among the most important applications of the computational analysis of vibrational spectra.

17.1 CALCULATION OF ALL FREQUENCIES OF ISOTOPIC SYSTEMS

It follows from theory that the force constant matrices for a molecule and all of its isotopic compositions are equal. Therefore the vibrational frequencies of all isotopic systems can be calculated if the force constant matrix for the molecule is known. The accuracy of the result depends of course on the harmonic oscillator model. It is therefore advisable to introduce anharmonicity corrections, especially for light atoms such as hydrogen, to increase the accuracy [99, 100].

Example 17.1   Vibrational frequencies of $O^{13}CO$ from the data for $O^{12}CO$ [101]
The calculation is based on the equality of force constants of isotopic compounds (Eq. (9.3.6)) and the equations of motion for the system YXY ($\mathbf{D}_{\infty h}$) (Eqs. (7.2.1) to (7.2.3)). The symmetrical stretching frequency $v_s = 1355$ cm^{-1} (Ra) is the same for $O^{12}CO$ and $O^{13}CO$ since the mass $m_Y - m_0$ is not altered. The asymmetrical stretching frequency v'_{as} and the deformation frequency v'_α of the $O^{13}CO$ molecule can be calculated according to

$$v'_{as \, or \, \alpha} = \sqrt{\frac{\mu_Y + 2\mu_{X'}}{\mu_Y + 2\mu_X}} \times v_{as \, or \, \alpha} = 0.9716 \, v_{as \, or \, \alpha} \tag{17.1.1}$$

with

$$m_0 = 15.999; \quad m_X = m_{12_C} = 12.00 \quad \text{and} \quad m_{X'} = m_{13_C} = 13.00$$

and

$$\nu_{as} = 2396 \text{ cm}^{-1}; \quad \nu_\alpha = 673 \text{ cm}^{-1}(\text{IR}) \quad \text{for } O^{12}CO$$

it follows that

$$\nu'_{as} = 2328 \text{ cm}^{-1}; \quad \nu'_\alpha = 654 \text{ cm}^{-1} \text{ (IR)} \quad \text{for } O^{13}CO$$

The two frequencies ν'_{as} and ν'_α agree exactly with the experimental results although anharmonicity corrections had not been included in the calculations (cf. Example 1.3).

The equation for the calculation of the isotope frequencies for diatomic molecules can be derived from Eqs. (7.1.1) and (9.3.6):

$$\nu'_{XY} = \sqrt{\frac{\mu'_{X'} + \mu'_{Y'}}{\mu_Y + \mu_X}} \times \nu_{XY} \tag{17.1.2}$$

This allows the vibrational frequency to be calculated for, e.g., $\nu'_{XY} = \nu_{13_O}$ of ^{13}CO from the experimental data, $\nu_{XY} = \nu_{^{12}C} = 2143 \text{ cm}^{-1}$ to give $\nu_{^{13}C} = 2096 \text{ cm}^{-1}$, which is in excellent agreement with the experimental data [101, 102].

Both results show that the assignment of the vibrational frequencies of isotopic molecules can be made with great accuracy on the basis of model calculations (cf. Example 17.3).

Example 17.2   The calculation of the frequencies of $^{14}N^{13}C^{34}S^-$ ($C_{\infty v}$) from complete data sets and by iteration from the data for isotopic compounds

The stretching frequencies $\nu_1(^{14}N^{13}C)$ and $\nu_2(^{13}C^{34}S)$ have to be calculated from the data for $^{14}N^{12}C^{32}S^-$ and $^{14}N^{13}C^{34}S^-$ by using the masses $m_{^{14}N} = 14.00$; $m_{^{12}C} = 12.00$; $m_{^{13}C} = 13.00$; $m_{^{32}S} = 31.97$; $m_{^{34}S} = 33.97$. The experimental stretching frequencies [103] (IR) are

$$^{14}N^{12}C^{32}S : \nu(NC) = 2086; \quad \nu(CS) = 757$$

$$^{14}N^{13}C^{34}S : \nu(NC) = 2037; \quad \nu(CS) = 751 \text{ cm}^{-1}$$

The **G** matrix elements are, according to Eq. (11.10),

NCS$^-$	g_{11}(NC)	g_{22}(CS)	g_{12}(NC/CS)
$^{14}N^{12}C^{32}S^-$	0.1548	0.1146	-0.08333
$^{14}N^{13}C^{32}S^-$	0.1483	0.1082	-0.07691
$^{14}N^{13}C^{34}S^-$	0.1483	0.1063	-0.07691

The calculation of the **F** matrix for the two stretching frequencies according to Eqs. (13.1)–(13.21) for complete sets of isotope data leads, according to Eq. (13.22), to

$$f_{NC} = 16.01; \quad f_{CS} = 5.06 \quad \text{and} \quad f_{NC/CS} = 0.94 \text{ N/cm}$$

The values are used together with the **G** matrix elements for $^{14}N^{13}C^{34}S^{-}$ in Eqs. (13.1) and (13.2) [104] to obtain the desired vibrational frequencies for $^{14}N^{13}C^{34}S^{-}$, which are v_1 ($^{14}N^{13}C$) = 2037 cm^{-1} and v_2($^{13}C^{34}S$) = 740 cm^{-1}. The experimental values are 2036 and 742 cm^{-1} respectively.

The calculation of the vibrational frequencies of isotopic compounds is of great practical interest, but the low orders $n = 2$ and 3 together with complete sets of isotope data are only realized for a few systems. The force constants that can be obtained with the data from the system $^{14}N^{13}C^{32}S$ by the iterative method described in Chapter 14,

$$f_{NC} = 16.16; \quad f_{CS} = 5.03 \quad \text{and} \quad f_{NC/CS} = 1.06 \text{ N/cm}$$

lead to the same vibrational frequencies as the calculations using a complete data set.

To summarize, it can be said that the isotope frequencies for problems of the orders $n = 2$ and 3 can be calculated with an accuracy of 2 to 3 digits. For higher orders a reduced accuracy is expected.

Example 17.3   Assignment of a shoulder in the region of the gold–chlorine stretching frequencies of AuCl(P(C$_6$H$_5$)$_3$)

Coates and Parkin [105] have assigned the gold–chlorine stretching frequency of triphenylphosphinegold(I) chloride to the intense infrared band at 329 cm^{-1} which shows a shoulder at 323 cm^{-1}. A simple estimation can clarify the assignment of this shoulder without the need to carry out a complete normal co-ordinate analysis with various force constant calculations. By using:

$$m_{Au} = 197; \quad m_{^{35}Cl} = 34.97; \quad m_{^{37}Cl} = 36.96; \quad v_{^{35}Cl} = 329 \text{ cm}^{-1}$$

in Eq. (17.1.2), 321 cm^{-1} is found as an approximate value for the Au ^{37}Cl frequency. This illustrates furthermore the observability of isotope frequencies in isotopic mixtures since the ratio of the concentrations of ^{37}Cl to ^{35}Cl is equal to 1:3 (cf. section 2.1.3).

17.2   CALCULATIONS OF SOME FREQUENCY REGIONS FOR STRONGLY COUPLED MOLECULAR SYSTEMS

It is sometimes possible to use force constant calculations to obtain frequency regions for some vibrations for a molecular system with strong vibrational coupling, even if no spectra of isotopic molecules are available.

Example 17.4   Upper limit for the internal $O-O$ stretching frequency of the isolated $Re_2\square O_{12}$ block of a perovskite

The $Re_2\square O_{12}$ unit in perovskites[†],[‡] of the type $Ba_4B^{II}(Re_2\square O_{12})$ can be treated as an isolated unit in an approximation of their vibrational properties. The unit consists of two ReO_6 octahedra which are joined by sharing faces with an octahedral vacancy in a linear arrangement. The **GF** matrix method for this unit with D_{3d} symmetry leads, for the species A_{1g}, to a very strong coupling between the $O-O$ bond and the neighbouring $Re-O$ bond. Reasonable coupling force constants are obtained only for an upper limit of $120-130$ cm^{-1} for the $O-O$ stretching frequency. In this way it was possible to assign the experimental frequencies that were observed around $100-115$ cm^{-1} to this internal $O-O$ stretching frequency which lies in the region of lattice vibrations (cf. section 21.3).

The external $O-O$ stretching frequencies of the ReO_6 units fall in the region around 300 cm^{-1}.

The $O-O$ stretching frequencies of the species A_{1g} are observed only in the Raman spectrum because of the centre of inversion of the $Re_2\square O_{12}$ unit (cf. Examples 1.3, 6.1, 6.2, and 6.4).

17.3 CONFIRMATION OF THE ASSIGNMENT OF EXPERIMENTAL FREQUENCIES

The confirmation of the assignments of frequencies to bonds or angles of a molecule is of great practical importance. We often proceed according to the following scheme: initially as many as possible of the frequencies are assigned by comparison with small molecules. The possible assignments of the remaining frequencies are tested with the **GF** matrix method. Wrong assignments lead to force constants which are either too large or too small, so that the choice of possible assignments can be narrowed.

Example 17.5   Selecting the $S-Cl$ stretching vibration in the infrared spectrum of SO_3Cl^- [106]

The $S-Cl$ stretching vibration was assigned to the band at 540 cm^{-1} following earlier assignments. This leads to an extremal force constant $f_{SCl} = 4.39$ N/cm by the stepwise coupling method (cf. Chapter 16). This is much higher than comparable valence force constants.

$$(f_{SCl}(SO_2Cl_2) = 2.5 \text{ N/cm}; \quad f_{SCl}(SCl_2) = 2.68 \text{ N/cm } [107])$$

Therefore the assignment $v_{SCL} = 416$ cm^{-1} was used in another calculation which gave the extremal force constant $f_{SCl} = 2.75$ N/cm. This assignment puts the valence force constant $f_{SCl}(SO_3Cl^-)$ in the range of previously observed values.

---

† This example illustrates the applicability of the methods for free molecules to isolated units in solids, as will be described in detail in Chapters 21 to 24.

‡ The extensive derivations can be found in the original literature [78].

Additional interpolations according to Siebert's rule (Chapter 20) give $f_{SCl} = 2.69$ N/cm and by Badger's rule (Chapter 19) give $f_{SCl} = 2.58$ N/cm. The assignment $v_{SCl} = 416$ cm^{-1} can therefore be regarded as being confirmed.

18

Trends in homologous compounds

Numerous trends can be observed if force constants are available for a series of homologous compounds (see also Table A.1).

18.1   INFLUENCE OF MOLECULAR SUBSTITUENTS

The exact calculation of the valence force constants in Example 8.1 has demonstrated the influence of substituents on the N$-$N bond of the diazoalkanes $NNCH_2$ and $NNC(Sn(CH_3)_3)_2$. Similar observations have been made for numerous azides NNN$-$R, where the valence force constant $f_{\beta N_\gamma N}$ of the nitrogen–nitrogen bond $\beta N-\gamma N$ does not increase in proportion to the frequency [52].

18.2   HYBRIDIZATION

The hybridization of C$-$H compounds gives rise to an increase in the valence force constant $f(CH)$ with increasing s-character [108] (e.g. from 4.95 N/cm for CH_4 (sp^3) to 5.90 N/cm for C_2H_2 (sp)).

18.3   IONIC CHARGE

For numerous series of compounds the valence frequencies and force constants increase with the ionic charge. Some diatomic compounds are listed as examples in Table 18.1.

Table 18.1 — Frequencies and force constants of O_2, O_2^-, NO^+ and NO

Compound	Frequency cm^{-1} [109]	Force constant N/cm
O_2	1556 (Ra)	11.41
O_2^-	1145 (Ra)	6.18
NO^+	2387 (Ra)	25.06
NO	1876 (Ra)	15.48

However this trend does not appear in all series [110].

18.4 CO-ORDINATION NUMBER

In many series the valence frequency is observed to decrease with increasing co-ordination number, e.g. in the series $(Ag(CN)_2)^-$ $v_{CN} = 2135$ cm^{-1}, $(Ag(CN)_3)^{2-}$ $v_{CN} = 2105$ cm^{-1} and $(Ag(CN)_4)^{3-}$ $v_{CN} = 2092$ cm^{-1} (IR) [111].

The influences of oxidation state, co-ordination number and charge on the vibrational frequencies, force constants and bond orders have also been studied. For halogen fluorides, for example, a decrease of bond strength has been found for increasing co-ordination number and oxidation number and constant charge [112].

18.5 TRENDS IN THE PERIODIC TABLE

The force constants that are listed in Table A.1 for the valences $X-X$, $H-X$, $C-X$, $N-X$, and $O-X$ with arbitrary atoms X clearly show the bonding properties in the periodic table of the elements. Numerous parallels to the bond energies can be drawn and more precise studies also reveal typical deviations [113] (see also wavenumber trends in Fig. 3.1).

19

Estimation of equilibrium distances by interpolation — Badger's method

The valence force constants can be used to calculate bond lengths if the force constants and bond lengths of simpler compounds are known. This is illustrated by using Badger's method:

Badger's rule [114,115] connects the bond length r_{XY} and the valence force constant f_{XY} of the bond as follows:

$$f_{XY} = \frac{c_{XY}}{(r_{XY} - d_{XY})^3}$$ (19.1)

or

$$r_{XY} = \left(\frac{c_{XY}}{f_{XY}}\right)^{1/3} + d_{XY}$$ (19.2)

The constants c_{XY} and d_{XY} are determined from the data for two similar molecules.

Example 19.1   Bond length estimation for ONF

The force constants f_{NO} and f_{NF} and the bond lengths and force constants of the three molecules O_2, N_2 and F_2 are available for the estimation of the bond lengths r_{ON} and r_{NF} of the triatomic molecule ONF (Table 19.1).

Table 19.1 — Bond lengths and force constants of
O_2, N_2, F_2 and NOF

Molecule	r (nm)	f (N/cm)
O_2 [116]	0.1207	11.41
N_2	0.1100	22.39
F_2	0.1417	4.45
NOF [117]		
ON		15.93
NF		2.22

Step 1:   The system-specific constants c_{XY} and d_{XY} are determined from the data of the homonuclear compounds X_2 and Y_2 according to

$$d_{XY} = \frac{f_{XX}^{1/3} r_{XX} - f_{YY}^{1/3} r_{YY}}{f_{XX}^{1/3} - f_{YY}^{1/3}} \qquad (19.3)$$

and

$$c_{XY} = f_{XX} (r_{XX} - d_{XY})^3 \qquad (19.3a)$$

Step 2: The bond lengths are calculated according to

$$r_{XY} = r_{ON} = \left(\frac{c_{ON}}{f_{ON}}\right)^{1/3} + d_{ON} \qquad (19.4)$$

and

$$r_{XY} = r_{NF} = \left(\frac{c_{NF}}{f_{NF}}\right)^{1/3} + d_{NF} \qquad (19.5)$$

Result: $r_{ON} = 0.115$ nm and $r_{NF} = 0.162$ nm.
Experimental values: $r_{ON} = 0.113$ nm and $r_{NF} = 0.152$ nm [118].

20

Determination of bond orders — Siebert's method

The bond order N can be estimated in many cases directly from the force constant f_{XY}. The bond order is defined according to Siebert [119–121] as the quotient of the restoring force $K_{XY,N}$ of the multiple bond and the restoring force $K_{XY,1}$ of the ideal single bond. This ratio can be expressed in terms of force constants and bond lengths:

$$N = \frac{K_{XY,N}}{K_{XY,1}} = \frac{f_{XY,N} r_{XY,N}}{f_{XY,1} r_{XY,1}} \tag{20.1}$$

$f_{XY,N}$ is the force constant which has been determined from the vibrational spectrum and $r_{XY,N}$ is the bond length of the substance that is under investigation. $r_{XY,1}$ can be approximated by

$$r_{XY,1} = r_{X,1} + r_{Y,1} = 0.06\left(\frac{n_X^3}{Z_X} + \frac{n_Y^3}{Z_Y}\right) \text{ nm} \tag{20.2}$$

where $r_{X,1}$ and $r_{Y,1}$ are the covalent radii according to Pauling [122], Z_X and Z_Y are the atomic numbers, and n_X and n_Y are the main quantum numbers of the valence electrons of the atoms X and Y.

By starting from methylene compounds and theoretical considerations, Siebert has constructed a formula for calculating the force constant of the ideal single bond:

$$f_{XY,1} = 7.20\frac{Z_X Z_Y}{n_X^3 n_Y^3} \quad \text{(N/cm)} \tag{20.3}$$

The factor 7.20 is a result of empirical investigations. If $r_{XY,N}$ and $r_{XY,1}$ are not known† [123] a further approximation can be obtained from the empirical relationship between f and r:

$$N = 0.57\frac{f_{XY,N}}{f_{XY,1}} + 0.43 \sqrt{\frac{f_{XY,N}}{f_{XY,1}}} \tag{20.4}$$

This simplified interpolation requires for the estimation of the bond order only the corresponding force constant, the atomic number of the atoms involved and the main quantum number of their valence electrons.

Example 20.1   BrCN: bond order of the CN bond
Data:

$$f_{CN,N} \qquad = 18.1 \text{ N/cm } [86]; \qquad r_{CN,N} = 0.1158 \text{ nm } [124]$$

$$r_{CN,1} \qquad = r_{C,1} + r_{N,1} = 0.0772 + 0.070 = 0.1472 \text{ nm } [125]$$

$$f_{CN,1} \qquad = \frac{7.2 \times 7 \times 6}{2^3 \times 2^3} = 4.725 \text{ N/cm.}$$

Results:
With the known bond length $r_{CN,N}$ (Eq. (20.1)):
$N = 3.01$ ($N = 2.99$ with $r_{CN,1} = 0.1486$ nm according to Eq. (20.2));
without $r_{CN,N}$ (Eq. (20.4)): $N = 3.03$.
Comparison of the results from Eqs. (20.1) and (20.4) shows that Eq. (20.4) should be sufficiently accurate for many applications.

Bond orders and mesomeric structures
The weighting of mesomeric structures can be estimated from bond order calculations which yield non-integer bond orders. In the case of NNO, the bond orders $N_{NN} = 2.6$ and $N_{NO} = 1.6$ indicate that the mesomeric structures $N\equiv N-O$ and $N=N=O$ are equally represented. In the case of $(CNO)^-$ with $N_{CN} = 2.7$ and $N_{NO} = 1.2$ ($C\equiv N-O$) dominates over ($C=N=O$) [126].

Remarks
The above interpolations have been used in numerous cases to determine bond orders and to solve molecular structures [127].‡ However, in some cases difficulties are encountered. In most cases integer values can be interpreted as single, double or triple bonds and intermediate values as a combination of mesomeric structures. In this respect the bond order has a real structural meaning. Since (apart from some transition metal compounds) the highest multiplicity is 3, bond orders up to about $N = 3$ appear to be sensible. Therefore the maximum value of the corresponding

---

† The bond length can be found by an iterative method which is based on Badger's rule as modified by Jenšovský [123].
‡ A compilation of bond orders according to Siebert and molecular orbital calculation can be found in [127]. This shows a relatively good agreement of the data from the two methods.

force constant is expected to be about $f_{max} = f_{XY,3} \approx 3.8\, f_{XY,1}$.[†]

Initially the N_2 molecule was the only known exception, with $N = 3.14$ [119]. Since then, several nitrogen–sulphur compounds have been synthesized for which the interpolation yields larger values (cf. Table 20.1).

A meaningful interpretation of these results in a structural context is not possible. Therefore a correction has been proposed for the nitrogen–sulphur system which has been taken into account in N_{NS} (corr.) in Table 20.1 [128].

Table 20.1 — Vibrational frequencies (IR, Ra), force constants and bond orders of N−S bonds

Molecule	$v_{NS}[cm^{-1}]$	$f_{NS}[N/cm]$	N_{NS}	$N_{NS}(corr.)$
$F_5As–NSF_3$	1610	14.88	3.1	2.7
$[Re(CO)_5NSF_3]^+$	1643	15.77	3.3	2.8
CH_3NSF_3	1788	17.53	3.6	3.0

Finally we would like to point out that there is a compilation of bond orders of X−X, H−X, C−X, N−X and O−X compounds in the Appendix. This illustrates the usefulness of Siebert's method for the calculation of bond orders.

[†] From Eq. (20.1) and $r_{XY,3} = 0.79\, r_{XY,1}$ [119] it follows that $3f_{XY,1}r_{XY,1} = f_{XY,3} \times 0.79\, r_{XY,1}$, so that $f_{XY,3} \approx 3.8\, f_{XY,1}$.

Part V
Vibrational spectroscopy of solids

The vibrational spectroscopy of crystalline solids is an important area which helps to clarify many problems relating to structure and bonding. In many cases the methods of vibrational spectroscopy have proved to be superior to other spectroscopic methods.

Chapter 21 summarizes some important results of the theoretical and experimental methods of the vibrational spectroscopy of solids.

Chapter 22 deals with the fundamentals of the group theory of crystalline solids and introduces factor group analysis.

Chapters 23 and 24 are dedicated entirely to the important areas of high-pressure studies and the investigation of molecules on the surface of metal crystals.

21

Introduction to spectroscopy of solids

A brief introduction to the vibrational spectroscopy of solids as an extension of molecular vibrational spectroscopy will be followed by a classification scheme for crystalline solids, including their vibrational characteristics, and a discussion of the differentiation of molecular and lattice vibrations. Introductory literature is available [43, 129–131].

21.1 MATERIAL ADDITIONAL TO THAT FOR MOLECULES

Translational and rotational motions have to be considered in the case of solids. These give rise to the lattice and acoustic vibrations of a solid.

Compared with molecules, the number of atoms (ions) has increased by several orders of magnitude. We will therefore consider only ordered solids with periodic lattices that can be characterized in spectroscopic terms by means of clear models and schemes.

21.2 CLASSIFICATION OF CRYSTALLINE SOLIDS BY 4 BOND TYPES — COUPLING OF MOLECULAR VIBRATIONS WITH LATTICE VIBRATIONS AND THE CASE OF SEPARABLE MOLECULAR VIBRATIONS

Crystals can be described roughly as a mixture of 4 bond types [32].

21.2.1 Molecular crystals

In molecular crystals, the molecules are kept in place by weak van-der-Waals forces (e.g. H_2, O_2). The molecules in the crystal lattice retain their molecular vibrations to a large extent, since the intramolecular forces are much stronger than the intermolecular forces.

The intramolecular and the lattice vibrations are well separated. The assignment of vibrations can draw on the extensive knowledge of the vibrations of free molecules.

21.2.2 Ionic crystals

Because of the strong electrostatic interactions between heteropolar ionic charges, we often find strong coupling between intramolecular and lattice vibrations. The assignment of vibrations may therefore require additional investigations.

However, a number of ionic crystals are known where the intramolecular and lattice vibrations are largely uncoupled, owing to system-specific conditions. In divanadates $X_2V_2O_7$, with X = Cd, Zn, Mg, Ba, Pb, Co, and Ni, it is possible for example to establish linear $V-O-V$ bridges spectroscopically. These system-specific separations arise from the distribution of charges and the co-ordination properties of the atoms involved.

21.2.3 Covalent crystals

The electrons in covalent bonds are shared between several atoms. The structure of covalent solids is dictated by the strengths and directions of the covalent bonds which may lead to less than optimum space-filling arrangements. A strong coupling of valence and lattice vibrations is caused by these highly directional and long-ranging valence forces.

Under certain circumstances sufficient separation between internal molecular and lattice vibrations occurs, and this then permits the identification and assignment of all vibrations. Among these systems are the spinels A_2BX_4, with the oxidation states of A and B of I–VI and II–IV [135] and the perovskites $A^{II}BMO_6$, with the oxidation states of the B and M ions of I–VII, II–VI and III–V [136] and A_4B^{II} $(M_2\square O_{12})$ with the oxidation states of the B and M ions of II–VI (M=W), II–VII (M=Re) [78] (cf. Example 17.4).

21.2.4 Metallic crystals

Metallic crystals are described by a model that assumes the complete delocalization of the valence electrons. The binding forces are non-directional, as in the case of the ionic crystals. Vibrational spectra cannot be obtained with conventional Raman or infrared spectrometers because of the strong absorption/reflection of electromagnetic radiation.

Exceptions are the spectra of surface molecules on metals which are discussed in detail in Chapter 24.

21.3 FREQUENCY REGIONS

The frequency regions of the vibrations of solids can be grouped according to the following criteria:

Molecular vibrations: the frequencies of the molecules (ions) in the crystal lattice correspond largely to those of the free molecules, i.e. they fall into the range $30\,\mathrm{cm}^{-1}$ [137] to $4000\,\mathrm{cm}$ [138]. However in most cases they start at around 200–300 cm.

Translatory lattice vibrations: these start in the range 50–$100\,\mathrm{cm}^{-1}$ and extend up to $300\,\mathrm{cm}^{-1}$, rarely beyond $400\,\mathrm{cm}^{-1}$.

Rotatory lattice vibrations: these fall in the range from 20 to 300 cm.

Acoustic vibrations: these are exactly equal to zero for a wave vector $k = 0$ and extend up to around $200\,\mathrm{cm}^{-1}$ for $k > 0$ [139].

The strongest couplings are to be expected between the molecular and the translatory lattice vibrations for molecular vibrations that lie in the region of the lattice vibrations and are of the same symmetry type.

21.4 DISCRIMINATION BETWEEN MOLECULAR AND LATTICE VIBRATIONS BY THEIR TEMPERATURE AND PRESSURE DEPENDENCE

The internal vibrations of molecules change very little with temperature. The lattice vibrations often show often a considerable increase in frequency with decreasing temperature.

The frequencies of lattice vibrations also show an increase at high pressures, whereas the frequencies of molecules in the lattice are much less affected [140–143].

These are two experimental methods that can distinguish between lattice and molecular vibrations in solids.

22

Group theory of crystalline solids

The fact that a crystalline solid is built out of unit cells causes the most significant reduction in the number of vibrations. Because of this, we need only study the vibrational behaviour of the unit cell with its relatively small number of vibrating particles.

The spectroscopic unit cell is usually more or less symmetric, so that numerous assignments can be made. Further reductions follow in the case of highly symmetric unit cells.

The small number of vibrations of crystalline solids is the spectroscopic indication of the usefulness of these concepts.

Introductory literature and tables are available [144–146].

22.1 SYMMETRY ELEMENTS, CRYSTAL CLASSES AND SPACE GROUPS OF CRYSTALLINE SOLIDS

An ideal, regular crystal lattice can be built up from the corresponding unit cell by translations in all directions of space. This adds the screw axis n_p and the glide plane c to the five symmetry elements of the free molecule, which are the identity I, the centre of symmetry i, the mirror plane σ, the axis of rotation C_p and the rotation inversion axis S_p. The screw axis n_p consists of a combination of a rotation of the order n with a translation by the fraction p/n of the unit cell. The orders are now restricted to $n = 2, 3, 4$ and 6 since the 5- and 7-fold axes of rotations are incompatible with crystalline order. For p it follows that $p = 1, 2, 3, \ldots, n\text{-}1$. The glide plane c is a combination of a translation and a reflection. The restriction of the order n leads to the 32 crystal classes or point groups. The unrestricted combination of symmetry elements leads to the 230 space groups which can be found in many texts [147–149]. The space group of a crystal can be determined experimentally by X-ray diffraction methods [150].

22.2 CRYSTALLOGRAPHIC AND SPECTROSCOPIC UNIT CELL

When a macroscopic crystal is broken up, the full symmetry is retained by the fragments. In principle this process can be continued until the smallest fragment that still retains the full symmetry of the macroscopic crystal is obtained. This is called the crystallographic unit cell.

Bravais came to the conclusion that all existing crystals can be constructed from 14 lattice types. These result from a combination of the 7 crystal systems with the 32 crystallographic point groups [151]. They are related to the crystallographic unit cell by translations to equivalent points in the unit cell and show the symmetry operations of the crystallographic unit cell.

The connection between the crystallographic and the spectroscopic unit cells is given by an integer multiplication:

If Z^S is the number of molecules in the spectroscopic unit cell, Z is the number of molecules in the crystallographic unit cell and LP is the number of lattice points that reduce the crystallographic unit cell to the spectroscopic unit cell [152]†, then

$$Z^S = \frac{Z}{LP} \tag{22.2.1}$$

The spectroscopic unit cell has proved to be a useful concept for the interpretation and assignment of vibrational spectra of crystalline solids by means of factor group analysis.

22.3 FACTOR AND SITE GROUPS

The translations of a unit cell are carried out in all directions in a space group. The translations of a unit cell in a factor group are restricted to a single translation (i.e. to two unit cells). The factor group of a space group is therefore a finite group. It is isomorphic with the 32 crystal classes corresponding to 32 point groups. The character tables of the factor groups follow therefore from the character tables of the isomorphic point groups.

In a factor group analysis all vibrations, the translatory and rotatory vibrations of the lattice and its components, the acoustic vibrations and the molecular vibrations of the molecules or ions (if present) in the unit cell are determined group theoretically from the spatial configuration of the components.

For each of the 230 space groups a unit cell can be constructed from a finite number of positions (lattice points, atoms, molecules). These are called sites if their symmetry is a sub-group of the space group of the crystal and the molecular point symmetry of the isolated molecule. The site symmetry for each lattice point consists of all symmetry elements of the lattice with respect to the lattice point excluding translation. The site groups as sub-groups of space and point groups of the unit cell play an important role in the correlation diagrams which are used to determine the number of vibrations in a unit cell (cf. Table 22.3).

22.4 FACTOR GROUP ANALYSIS

The factor group analysis can be carried out in three ways.

— By means of equations giving the reducible characters for the vibrational motions in the unit cell (analytical method).

---

† $LP = 1$ for primitive lattices (P); $LP = 3$ or 1 for trigonal lattices (R); $LP = 2$ for space-centred lattices (I); $LP = 2$ for base-centred lattices (A, B or C); $LP = 4$ for face-centred lattices (F).

— By means of the tabular method.
— By means of the correlation method using site groups.

All three methods are based on the spectroscopic unit cell.

22.4.1 Analytical method

The analytical method, which has been developed by Bhagavantam and Venkatar-ayudu [153], is a consequent extension of the group theoretical methods for the vibrational analysis of free molecules to crystalline solids.

p	= the number of atoms in the unit cell,
s	= the number of molecules in the unit cell,
v	= the number of monoatomic systems in the unit cell,
N_R	= the number of atoms that pass into themselves or their equivalents under the symmetry operation R of the class j,
$N_R(p)$	= the number of groups of atoms that pass into themselves or their equivalents under the symmetry operation R of the class j,
$N_R(s)$	= the number of molecules (ions) of which the centre of gravity remains unchanged under the symmetry operation R,
$N_R(s - v)$	= $N_R(s)$ minus the number v of monoatomic systems that remain unchanged under the symmetry operation R,
j	= class,
ϕ_j	= the angle of rotation associated with the symmetry operation R of the class j.

The reducible characters of the representation are

$$\chi_R(n) = N_R(p) \, (\pm 1 + 2 \cos \phi_j) \tag{22.4.1}$$

for all vibrations of the unit cell;

$$\chi_R(T) = \chi_R(\mathrm{ac}) = \pm 1 + 2 \cos \phi_j \tag{22.4.2}$$

for the translations of the unit cell as a whole, i.e. the acoustic vibrations;

$$\chi_R(T') = \chi_R(\mathrm{trans}) = [N_R(s) - 1] \, (\pm 1 + 2 \cos \phi_j) \tag{22.4.3}$$

for the translatory lattice vibrations;

$$\chi_R(R') = \chi_R(\mathrm{rot}) = N_R(s - v) \, (1 \pm 2 \cos \phi_j) \tag{22.4.4}$$

for the rotatory lattice vibrations†;

---

† This is valid for non-linear polyatomic groups, which is generally the case. The equations appropriate for linear polyatomic groups can be found in [154].

$$\chi_R(T') = \chi_R(\text{vib}) = [N_R - N_R(s)]\,(\pm 1 + 2\,\cos\,\phi_j) \tag{22.4.5}$$

$$- N_R(s - v)\,(+1 \pm 2\,\cos\,\phi_j)$$
$$= \chi_R(n) - \chi_R(ac) - \chi_R(\text{trans}) - \chi_R(\text{rot}) \tag{22.4.5a}$$

for the internal vibrations.

The numbers of the corresponding vibrations can be calculated with the group theoretical reduction formula (Eq. (5.5.1)) by using the appropriate character table for the space group of the solid.

Where

n	= the number of vibrations per symmetry species of the unit cell,
T	= the number of acoustic vibrations per symmetry species,
T'	= the number of translatory vibrations per symmetry species,
R'	= the number of rotatory vibrations per symmetry species,
n_i	= the number of internal vibrations of the molecules (ions) of the unit cell per symmetry species.

If the unit cell contains σ molecules, the number of optical vibrations in the case of molecular crystals is given by $3\sigma N{-}3$, which can be separated into $(3N{-}6)\sigma$ internal vibrations and $6\sigma{-}3$ lattice vibrations (N is the number of atoms in the molecule). Three vibrations remain for the acoustic vibrations of the crystal. This is illustrated in Fig. 22.1 as a function of the wave vector.

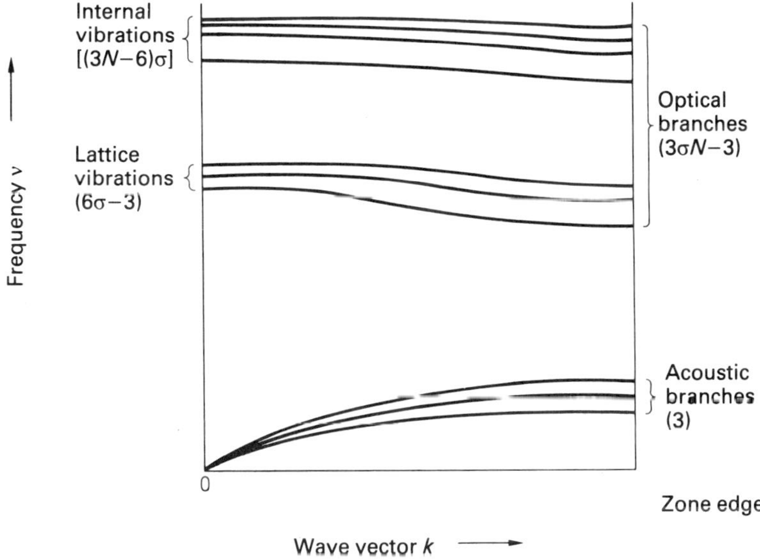

Fig. 22.1 — Schematic overview of the vibrations of a molecular crystal (taken from [155]).

Example 22.1 Factor group analysis for crystalline $CaCO_3$

The results of the factor group analysis for the calcite modification of $CaCO_3$ are shown in Table 22.1 (see also Example 6.5). Calcite has a rhombohedral structure, space group $\mathbf{D}_{3d}^6$ (No. 167). The spectroscopic unit cell contains two formula units $CaCO_3$, i.e. $Z^S = 2$ according to Eq. (22.2.1). The total number of vibrations is therefore 30 with 12 internal vibrations and 3 acoustic vibrations. There are 9 translatory and 6 rotatory lattice vibrations (cf. Table 22.1).

Table 22.1 — Character table and vibrations in the unit cell of $CaCO_3$

$\mathbf{D}_{3d}^6$	I	$2C_3$	$3C_2$	i	$2S_6$	$3\sigma_d$	n	T	T'	R'	n_i	IR	Ra
	Irreducible characters						Number of vibrations					Activity	
A_{1g}	1	1	1	1	1	1	1	0	0	0	1	ia	p
A_{1u}	1	1	1	-1	-1	-1	2	0	1	0	1	ia	ia
A_{2g}	1	1	-1	1	1	-1	3	0	1	1	1	ia	ia
A_{2u}	1	1	-1	-1	-1	1	4	1	1	1	1	a	ia
E_g	2	-1	0	2	-1	0	4	0	1	1	2	ia	dp
E_u	2	-1	0	-2	1	0	6	1	2	1	2	a	ia
							30	3	9	6	12		

Angle of rotation — number of atoms, molecules

	I	$2C_3$	$3C_2$	i	$2S_6$	$3\sigma_d$
ϕ_j (in°)	0	120	180	180	60	0
$N_R(p)$	10	4	4	2	2	0
$N_R(s)$	4	2	2	2	2	0
$N_R(s\text{-}v)$	2	0	2	0	0	0

reducible characters

	I	$2C_3$	$3C_2$	i	$2S_6$	$3\sigma_d$
$\chi_R(n)$	30	0	-4	-6	0	0
$\chi_R(T)$	3	0	-1	-3	0	1
$\chi_R(T')$	9	0	-1	-3	0	-1
$\chi_R(R')$	6	0	-2	0	0	0
$\chi_R(n_i)$	12	0	0	0	0	0

The observed vibrational spectrum can be interpreted in accordance with Table 22.1 and by using additional criteria [156]:

Internal vibrations [cm^{-1}]
$$n_i = \nu_{1g}(A_{1g}; \text{Ra}; CO: 1088) + \nu_{1u}(A_{1u}; \text{ia}; CO)$$
$$+ \nu_{2g}(A_{2g}; \text{ia}; OCO) + \nu_{2u}(A_{2u}; \text{IR}; OCO:871)$$
$$+ \nu_{3g}(E_g; \text{Ra}; CO: 1432) + \nu_{4g}(E_g; \text{Ra}; OCO: 714)$$
$$+ \nu_{3u}(E_u; \text{IR}; CO: 1410) + \nu_{4u}(E_u; \text{IR}; OCO: 712)$$

External vibrations:
$$n_{T'} = \nu(A_{1u}; \text{ia}) + \nu(A_{2g}; \text{ia}) + \nu(A_{2u}; \text{IR}: 303) + \nu(E_g; \text{Ra}: 156) + 2\nu(E_u; \text{IR}: 297$$
$$\text{and } 223)$$
$$n_{R'} = \nu(A_{2g}; \text{ia}) + \nu(A_{2u}; \text{ia}) + \nu(E_g; \text{Ra}: 283) + \nu(E_u; \text{IR}: 102)$$
$$n_{T=ac} = \nu(A_{2u}; \text{ia}) + \nu(E_u; \text{IR}: ?)$$

22.4.2 Tabular method

The tabular method of factor group analysis can be carried out by using the tables developed by Adams and Newton [146,157]. The following parameters are required: the space group of the crystal (number and symmetry), the number of molecules in the spectroscopic unit cell Z^S, and the symmetry characterization of the atoms or groups in Wyckoff nomenclature [158] (see Table 22.2).

Table 22.2 — Example: $CaCO_3$ (space group No. 167, $\mathbf{D}_{3d}^6$ ($R\bar{3}c$), $Z^S = 2$)

Number of atoms (Wyckoff notation)	A_{1g}	A_{2g}	E_g	A_{1u}	A_{2u}	E_u	Tables of Adams and Newton
2C (2A)		1	1		1	1	Table 2 (No. 167)
2Ca (2B)				1	1	2	Table 2
2·3O (6E)	1	2	3	1	2	3	Table 2
Vibrations							
n_t (total)	1	3	4	2	4	6	= 30 vibrations
$T' + T$ (2A + 2B)		1	1	1	2	3	Table 2
T (acoustic)					1	1	Character table
R' (2A)		1	1		1	1	Table 2
n_i (internal)	1	1	2	1	1	2	

The number of vibrations of the atoms C, Ca and 3O per symmetry species follows directly from the tables of Adams and Newton. Their sum gives the total number of vibrations. The number of translatory and acoustic vibrations per symmetry species $T' + T$ [caused by C and Ca (2A + 2B)] and the rotatory vibrations R' [caused by C (2A)] are also taken directly from the tables. The acoustic vibrations per symmetry species follow from the translations given in the character tables for the point group $\mathbf{D}_{3d}$. The internal vibrations n_i per symmetry species are obtained from $n_t - T' - T - R'$. The results are in complete agreement with those obtained by the analytical method†.

The tabular method is by far the fastest and most comfortable method for factor group analysis provided that the required data are available.

22.4.3 Correlation method

The correlation method is frequently used by chemists. It is easily applied to molecular crystals. Because a comprehensive description would take too much space we will only refer to the relevant literature [159].

As an example we show the correlation diagram (Table 22.3) for 2($CaCO_3$), space group No. 167 ($\mathbf{D}_{3d}^6$), which supplements and confirms the results of the first two methods.

---

† A number of other examples can be found in the tables of Adams and Newton [146] illustrating their use for the space groups 186, 136 and 164 for layered structures and polymers, and the construction of internal co-ordinates by using a vector table, as well as the procedure for point group analysis using the tables for factor group analysis.

Table 22.3 — Correlation diagram for 2(CaCO$_3$) space group 167 ($\mathbf{D}_{3d}^6$)

| Point group of free anion CO$_3^{2-}$ ($\mathbf{D}_{3h}$) | | Site group $\mathbf{D}_3$ | Factor of spectroscopic unit cell 2(CaCO$_3$) ($\mathbf{D}_{3d}^6$) | | | | Site group cation Ca^{2+} $S_6 = C_{3i}$ | |
Vibrations Internal	External	Species	Species	Vibrations of 2(CO$_3^{2-}$) Internal	External	External vibrations of 2 (Ca^{2+})	Species	Translatory vibrations
$2\nu_1$		A_1' → A_1 →	A_{1g}	ν_{1g}				
			A_{1u}	ν_{1u}		T_z		
	$2R_z$	A_2' → A_2 →	A_{2g}	ν_{2g}	R_z, T_z			
$2\nu_2$	$2T_z$	A_2'' →	A_{2u}	ν_{2u}	T_z, R_z	T_z	A_u	$2T_z$
$2\nu_3, 2\nu_4$	$2(T_x, T_y)$	E' → E →	E_g	ν_{3g}, ν_{4g}	$R_x, R_y; T_x, T_y$	$T_x, T_y;$		
	$2(R_x, R_y)$	E'' →	E_u	ν_{3u}, ν_{4u}	$T_x, T_y; R_x, R_y$	T'_x, T'_y	E_u	$2(T_x, T_y)$

23

High-pressure investigations

The vibrational frequencies of solids change at high pressures (cf. section 21.4). This effect can be used in structural investigations, some of which will be discussed in the following.

23.1 SYMMETRY CHANGES

The application of high pressures to crystalline solids often induces symmetry changes. The reduction of symmetry leads to the splitting of bands of degenerate vibrations. In $K_2[Zn(CN_4)]$, for example, the triply degenerate stretching vibration v_{CN} (F_2) of the $[Zn(CN)_4]^{2-}$ tetrahedron with a frequency of 2152 cm^{-1} at normal pressure splits successively into three infrared bands [160] (cf. section 6.2.1).

23.2 PHASE TRANSITIONS

Phase transitions are indicated by frequency jumps at characteristic high pressure. In KNO_2, for example, the jump of the deformation frequency of the NO_2^- ion from 808 to 820 cm^{-1} (IR) [161] at a pressure of 10 kbar indicates a phase transition.

Phase transitions may also show up as discontinuous removal of degeneracy of valence and deformation vibrations.

This is illustrated by the doubly degenerate v_{NO} (E') vibration of KNO_3 which is split into two bands at 1415 and 1360 cm^{-1} (IR) at 26 kbar [162].

23.3 IR SPECTROSCOPIC OBSERVATION OF HIGH PRESSURE SYNTHESIS

Some new compounds are only accessible by high-pressure synthesis. Such a compound is caesium diazoargentate $Cs[Ag(N_3)_2]$ which can be used without the hazard of explosion [163]. The synthesis from AgN_3 and CsN_3 at 10 kbar can be monitored by infrared spectroscopy, by using the symmetry reduction of the azide ion during complex formation.

23.4 PRESSURE-DEPENDENT REDUCTIONS

Infrared spectroscopy is probably the most convenient tool for studying pressure-dependent reductions, such as the reduction of Fe(III) to Fe(II).

The electron for the reduction

$$[\text{Fe(III)(NCS)}]_6^{3-} \ (\text{normal pressure}) \rightarrow [\text{Fe(II)(NCS)}]_6^{3-} \ (\text{high pressure})$$

comes from the ligand sphere [164]. The strong bands at 270 and 295 cm^{-1} have been assigned to the Fe(III)$-$N stretching vibration. At a pressure of 30 kbar these bands are shifted to 290 and 320 cm^{-1}, and an additional new band at 238 cm^{-1} appears which has been assigned to the Fe(II)$-$N stretching vibration. This indicates a strong decrease of the Fe$-$N bond strength in the high-pressure Fe(II) state.

23.5 REMOVAL OF THE JAHN–TELLER INSTABILITY

The removal of the Jahn–Teller instability[†] in octahedral Cu(II) complexes at pressures above 35 kbar can be observed by infrared spectroscopy [166].

The degeneracy of the corresponding stretching and deformation vibrations v_{PO} and v_{CoPO} of the compound $[(C_5H_5)Co(PO(OC_2H_5)_2)_3]_2Cu(II)$ increases with pressure, as is expected for a change of local symmetry from D_{4h} to O_h.

23.6 THE SHAPE-RESTORING FUNCTION OF THE DEFORMATION
FORCE CONSTANT

The change of bond angles is one of the effects that high pressure causes in solids. This can be shown, for example, in the infrared spectra of $NaNO_2$, where the bond angle in the NO_2^- ion is reduced from its normal value of 115.4° to 112° at 35 kbar. This can be calculated from the change of the deformation frequency from 820 to 833 cm^{-1} and the corresponding valence frequencies of the species A_1 by using Eqs. (13.1) to (13.21). It is interesting to note the increase in the deformation force constant f_{ONO} from 1.6 to 1.9 N/cm, which indicates its shape-restoring function [161][‡].

[†] A non-linear polyatomic system is, according to the Jahn–Teller theorem, stable only if its electronic ground state is non-degenerate. The symmetry must therefore be reduced until the degeneracy of the ground state has been removed completely.

[‡] Similar behaviour under normal pressures is found in the series
NO_2^+ ($\alpha = 180°$ and $f_\alpha = 0.47$ N/cm),
NO_2 ($\alpha = 134.3°$ and $f_\alpha = 1.10$ N/cm),
NO_2^- ($\alpha = 115.4°$ and $f_\alpha = 1.67$ N/cm),
where the decrease in the bond angle α is opposed by an increasing deformation force constant f_α [167,168].

24

Molecules on metal surfaces

Vibrational spectroscopy is ideally suited for the detection of interactions of molecules on the surfaces of metals and non-metallic solids. Experiments have shown that covalent bonds are formed only between surface molecules and the immediately neighbouring atoms of the solid. Thus, chemisorption can be described as an entirely local phenomenon. Only the surface molecule and the covalently bonded atoms of the solid have to be included in model calculations for which the computational methods described in Chapters 7 to 20 can be applied with minor adaptations. The translations and rotations of the surface molecules are now hindered and best described in the framework of solid-state vibrational spectroscopy (cf. section 21.1). The surface molecules can therefore be envisaged as a combination between free molecules and solids.

Special instrumentation is used to record Raman scattering from very small quantities of surface molecules which is enhanced by a factor of 100 to 100 000 through the process of chemisorption [169–173].

24.1 ACTIVATION OF INACTIVE VIBRATIONS BY SYMMETRY REDUCTION

Inactive vibrations of highly symmetric molecules are often activated by the reduced symmetry of the corresponding surface molecules. In the case of benzene C_6H_6 ($\mathbf{D_{6H}}$) the symmetry reduction after chemisorption activates eight vibrations that are inactive in the free molecule [173].

24.2 WEAK COUPLINGS IN BENZENE AND ITS ORIENTATION ON A METAL SURFACE

The vibrational frequencies of chemisorbed benzene differ by only a few cm^{-1} from those of gaseous benzene (mostly between 0 and $30\,cm^{-1}$, some by as much as 50–$70\,cm^{-1}$). This indicates that the electron configuration of benzene on the surface is largely retained.

Only the out-of-plane vibration $v(HCC_2)$ (A_{2u}) at $776\,cm^{-1}$ of benzene on Ni/SiO$_2$ differs by about $100\,cm^{-1}$ from that of free benzene ($671\,cm^{-1}$). This indicates that the benzene molecules are oriented parallel to the nickel surface [174].

24.3 STRONG INTERACTION IN THE CASE OF CARBON MONOXIDE

The Raman spectra and the electron energy loss spectra (EELS) of the system CO/Ni$-$SiO$_2$ have been measured in Raman cells with an internal pressure of 1 torr [175]. The vibrational frequency of free carbon monoxide is $2134\,cm^{-1}$. For CO chemisorbed onto Ni$-$SiO$_2$, frequencies at 2022, 2070, $2086\,cm^{-1}$, and $1804\,cm^{-1}$ (Ra) ($1848\,cm^{-1}$ EELS) have been observed. The first three vibrations have been interpreted as $v(CO)$ of OC$-$Ni. The band at $1804\,cm^{-1}$ is attributed to $v(CO)$ of OC$-$Ni$_4$. The C$-$Ni$_4$ stretching vibration of the linear OC$-$Ni surface molecule is assigned to the band at $355\,cm^{-1}$ and those of the bridged species OC$-$Ni$_4$ are assigned to the bands at 345, 362 and $640\,cm^{-1}$. Force constant calculations have been performed to support the assignment.

The strong decrease in the CO frequency from the free CO molecule to the OC$-$Ni$_4$ surface species indicates a strong alteration in the CO electron configuration.

24.4 METAL–METAL VIBRATIONS

By comparing the vibrational spectra of many molecule/Ni$-$SiO$_2$ systems it was possible to assign the band that was always present at $85\,cm^{-1}$ to the Ni$-$Ni vibration [176, 177]. The observation of metal–metal vibrations is facilitated by the enhanced Raman scattering of chemisorbed surface molecules.

Appendix

COMPILATION OF VALENCE FORCE CONSTANTS OF X−X, H−X, C−X, N−X, AND O−X BONDS FROM INORGANIC, ORGANIC AND SOLID-STATE CHEMISTRY

In Table A.1 valence force constants (N/cm) selected from various areas of chemistry are listed to provide a basis for further investigations.

The results obtained with complete sets of data are labelled c d. The other values have been obtained using various approximate methods (mostly the extremal methods discussed in Chapters 14 and 16).

The bond orders have all been calculated by using Siebert's method (Eqs. (20.1) and (20.4)); exceptions are labelled + (e.g. N−S).

The force constants in free molecules and solids are listed together to give a better overview.

Ranges of force constants of free molecules
The force constants of numerous free molecules that have been calculated so far fall into the following ranges:

Valence force constants: 0.1 to 26 N/cm
(for example Na_2: 0.17 N/cm and NO^+: 25.07 N/cm).

Deformation force constants: 0.01 to 2 N/cm
(for example HgI_2: 0.018 N/cm and NO_2^-: 1.72 N/cm).

The torsion force constants and the out-of-plane deformation force constants fall into the same range.

Coupling force constants can have values between −3 and +3 N/cm (for example $f_{ON/AF} = 2.5 \pm 0.36$ N/cm in ONF).

Ranges of force constants in solids

According to Schrader [19][†] the following ranges of force constants are observed in crystalline solids:

For bonds in molecules in crystal lattices: 0.1 to 30 N/cm.

For intermolecular forces in crystals: 0.001 to 0.4 N/cm.

For hydrogen bonds and ionic groups: 0.1 to 1 N/cm.

[†] References in this Appendix are given in the 'Literature for the Appendix'.

Table A.1 — Valence force constants and bond orders of X$-$X, H$-$X, C$-$X, N$-$X, and O$-$X bonds of inorganic, organic and solid compounds

Valence X$-$Y	Valence force constant f_{XY}	Bond order N	Calc. method	Compound	Literature [Ref.] (p. no.) {see p. 196}
H$-$H	5.14 (5.74)	0.77 0.84	c d anh.	H_2	[7](282)
Li$-$Li	1.24	1.2	c d	Li_2	[20] (501)
B$-$B	3.58	1.2	c d	B_2	[13] (107)
C$-$C	16.5	3.2	c d	HCCH (see also under C$-$X)	[7] (290)
N$-$N	22.42	3.2		N_2 (see also under N$-$X)	
O$-$O	11.41	1.4		O_2 (see also under O$-$X)	
F$-$F	4.45	0.58	c d	F_2	[7] (282)
Na$-$Na	0.17	0.24	c d	Na_2	[7] (283)
Si$-$Si	4.65	2.0	c d	Si_2	[13] (107)
Si$-$Si	≈ 1.7	≈ 0.9		Si_2H_6	[1] (172)
P$-$P	5.56	2.1	c d	P_2	[13] (107)
P$-$P	2.07	0.95	c d	P_4	[7] (284)
S$-$S	4.96	1.7	c d	S_2	[13] (108)
S$-$S	2.5	0.99		S_8	[4] (282)
Cl$-$Cl	3.24	1.1	c d	Cl_2	[7] (282)
Ni$-$Ni	0.1	0.2		solid Ni	[9]
As$-$As	3.91	1.8	c d	As_2	[13] (109)
Se$-$Se	3.61	1.6	c d	$^{80}Se_2$	[13] (108)
Br$-$Br	2.36	1.1	c d	Br_2	[7] (282)
Rb$-$Rb	0.08	0.2	c d	Rb_2	[13] (106)
Cd$-$Cd	1.11	1.0	c d	Cd_2^{2+}	[4] (40)
Sb$-$Sb	2.61	1.9	c d	Sb_2	[13] (108)
Te$-$Te	2.37	1.7	c d	Te_2	[13] (108)
I$-$I	1.70	1.2	c d	I_2	[7] (282)
Hg$-$Hg	1.69	1.5	c d	Hg_2^{2+}	[4] (40)
Pb$-$Pb	4.02	3	c d	Pb_2	[13] (107)
Bi$-$Bi	1.84	1.6	c d	Bi_2	[13] (108)

Valence X−Y	Valence force constant f_{XY}	Bond order N	Calc. method	Compound	Literature [Ref.] (p. no.) {see p. 196}
H−H	5.14 (5.74)	0.77 0.84	c d anh.	H_2	[7] (282)
H−B	2.75	0.68	c d	BH_3	[4] (65)
H−C	5.50	1.0	c d	CH_4	[8] (44)
H−N	7.05	1.1	c d	NH_3	[8] (44)
H−O	8.45	1.1	c d	H_2O	[7] (289)
H−O	7.40	1.0	c d	HO^-	[7] (282)
H−F	8.85	1.1	c d	HF	[7] (282)
H−Al	1.76	0.60	c d	AlH_4^-	[4] (65)
H−Si	2.98	0.85	c d	SiH_4	[7] (293)
H−P	3.11	0.82	c d	PH_3	[7] (293)
H−S	4.29	1.0	c d	H_2S	[7] (289)
H−Cl	4.81	1.0	c d	HCl	[7] (282)
H−Ge	2.81	0.82	c d	GeH_4	[7] (293)
H−As	2.85	0.81	c d	AsH_3	[8] (44)
H−Se	3.51	0.93	c d	H_2Se	[7] (289)
H−Br	3.84	0.98	c d	HBr	[7] (282)
H−Sn	2.03	0.76	c d	SnH_3	[4] (65)
H−Sb	2.09	0.77	c d	SbH_3	[7] (292)
H−I	2.92	0.97	c d	HI	[7] (283)
C−H	5.50	1.0	c d	CH_4	[8] (44)
C−B	3.82	1.1		$B(CH_3)_3$	[21]
C−C	16.5	3.2	c d	HCCH	[7] (290)
C−C	9.15	1.9		H_2CCH_2	[5] (200)
C−C	7.6	1.7		C_6H_6	[3] (300)
C−C	4.4	1.1		H_3CCH_3	[5] (185)
C−N	18.07	3.0	c d	HCN	[10] (32)
C−N	11.84	2.1	c d	CN_2^{2-}	[7] (284)
C−N	6.54	1.3		$NNCH_2$	[11]
C−O	18.56	2.8	c d	CO	[7] (283)
C−O	15.61	2.4	c d	CO_2	[7] (283)
C−O	12.76	2.0	c d	OCH_2	[7] (296)
C−O	7.86	1.3	c d	CO_3^{2-}	[7] (291)
C−O	5.1	0.96		$O(CH_3)_2$	[5] (198)
C−F	6.98	1.1	c d	CF_4	[7] (293)
C−P	8.95	2.4	c d	HCP	[7] (289)
C−S	7.67	2.6	c d	CS_2	[7] (283)
C−S	3.3	1.0		$S(CH_3)_2$	[5] (198)
C−Cl	3.12	0.93	c d	CCl_4	[7] (293)
C−Ni	2.91	1.2		Ni_4CO	[9] (22)
C−Ni	1.43	0.68		NiCO	[9] (22)
C−Se	5.94	1.8		CSe_2	[7] (283)
C−Br	2.42	0.86	c d	CBr_4	[7] (293)
C−Rh	2.4	1.2		$[Rh(CN)_6]^{3-}$	[4] (152)
C−Ag	2.0	0.99		$[Ag(CN)_2]^-$	[4] (152)
C−I	1.69	0.79	c d	CI_4	[7] (293)

Valence X−Y	Valence force constant f_{XY}	Bond order N	Calc. method	Compound	Literature [Ref.] (p. no.) {see p. 196}
N−H	7.05	1.1	c d	NH_3	[8] (44)
N−B	7.2	1.6	c d	BN_2^{3-}	[7] (284)
N−B	5.5	1.3		$B(N(CH_3)_2)_3$	[4] (134)
N−C	18.07	3.0	c d	HCN (cf. Table 7.1)	[10] (32)
N−N	22.42	3.2	c d	N_2	[7] (282)
N−N	18.09	2.6	c d	$NNCH_2$ (cf. Example 8.1)	[11]
N−N	16.0	2.4	c d	N−NNH	[23]
N−N	13.15	2.0	c d	$N-N-N^-$	[7] (284)
N−O	25.07	3.1	c d	NO^+	[7] (283)
N−O	17.17	2.3	c d	NO_2^+	[7] (283)
N−O	15.49	2.1	c d	NO	[7] (283)
N−O	15.18	2.0	c d	ONCl	[7] (295)
N−O	11.78	1.7	c d	NNO (cf. Example 13.1)	
N−F	4.16	0.66	c d	NF_3	[7] (292)
N−F	3.01	0.52	c d	ONF	[7] (295)
N−Si	3.8	1.1		$((CH_3)_3Si)_2NH$	[4] (132)
N−Si	4.8	1.4		H_3SiNCS	[4] (126)
N−P	6.7	1.7		$P_3N_3Cl_6$	[4] (130)
N−S	17.53	3.0^+		$CH_3NSF_3^+$ (cf. Table 20.1)	[12] (89), [17]
N−S	14.9	2.7^+		F_5AsNSF_3 (cf. Table 20.1)	[12] (89), [17]
N−S	12.54	2.5^+		NSF_3	[12](88), [17]
N−S	10.7	2.2^+		NSF_3	[4](52), [17]
N−S	8.3	1.9		HNSO	[18]
N−S	6.2	1.5		$S_3N_3O_3Cl_3$	[18]
N−S	3.1	0.87		H_3N-SO_3	[18]
N−Cl	2.39	0.68	c d	ONCl	[7] (295)
N−Br	1.13	0.43	c d	ONBr	[7] (295)
O−H	8.45	1.1	c d	H_2O	[7] (289)
O−Li	1.58	0.66	c d	LiO	[4] (41)
O−Be	7.51	1.8	c d	BeO	[13] (106)
O−B	13.66	2.5	c d	BO	[13] (107)
O−B	11.4	2.1	c d	BO_2^-	[7] (283)
O−B	6.35	1.3	c d	BO_3^{3-}	[7] (291)
O−C	18.56	2.8	c d	CO (cf. C−O)	[7] (283)
O−N	25.07	3.1	c d	NO^+ (cf. N−O)	[7] (283)
O−O	16.59	2.0	c d	O_2^+	[13] (108)
O−O	11.41	1.4	c d	O_2	[4] (40)
O−O	6.18	0.89	c d	O_2^-	[4] (40)
O−O	5.70	0.83	c d	O_3	[7] (289)
O−O	0.71	0.19		$Ba_4Cd(Re_2\square O_{12})$ internal O−O	[14]
O−O	0.55	0.16		external O−O	[14]
O−F	3.95	0.58	c d	OF_2	[7] (289)
O−Na	≈3.2	≈1.1		NaOH	[24] (142, 180)
O−Mg	3.5	1.1	c d	MgO	[13] (106)
O−Al	5.66	1.5	c d	AlO	[13] (107)
O−Al	3.8	1.1		$[Al(OH)_4]^-$	[4] (145)
O−Si	9.25	2.1	c d	SiO	[13] (107)
O−Si	4.75	1.2	c d	SiO_4^{4-}	[4] (68)
O−P	9.41	2.0	c d	PO	[13] (108)
O−P	6.16	1.4		PO_4^{3-}	[4](68)
O−S	10.01	2.0	c d	SO_2	[7] (289)
O−S	7.93	1.6	c d	SO	[13] (108)
O−Cl	7.02	1.5	c d	ClO_2	[7] (289)
O−Cl	4.26	1.0		ClO_2^-	[4] (50)
O−Cl	3.30	0.82	c d	ClO^-	[7] (283)

Valence X−Y	Valence force constant f_{XY}	Bond order N	Calc. method	Compound	Literature [Ref.] (p. no.) {see p. 196}
O−Ca	2.85	1.2	c d	CaO	[13] (107)
O−Sc	6.56	2.3	c d	ScO	[13] (108)
O−Ti	7.19	2.4	c d	TiO	[13] (108)
O−V	7.36	2.3	c d	VO	[13] (108)
O−Cr	5.82	1.9	c d	CrO	[13] (108)
O−Mn	5.16	1.6	c d	MnO	[13] (108)
O−Fe	5.67	1.7	c d	FeO	[13] (108)
O−Fe	4.72	1.5		FeO_4^{2-}	[4] (68)
O−Cu	2.97	0.93	c d	CuO	[13] (108)
O−Ga	4.52	1.2	c d	GaO	[13] (108)
O−Ge	7.53	1.8	c d	^{74}GeO	[13] (107)
O−As	7.27	1.7	c d	AsO	[13] (108)
O−Se	6.45	1.5	c d	SeO	[13] (108)
O−Se	4.55	1.1		SeO_3^{2-}	[4] (58)
O−Br	5.28	1.3		BrO_3	[4] (58)
O−Sr	3.40	1.4	c d	SrO	[13] (107)
O−Y	5.81	2.2	c d	^{89}YO	[13] (108)
O−Zr	7.38	2.6	c d	ZrO	[13] (109)
O−Nb	2.8	1.1		Ba_2GdNbO_6 (solid)	[15]
O−Mo	3.05	1.2		Ba_2CaMoO_6 (solid)	[15]
O−Ru	6.70	2.2	c d	RuO_4	[7] (294)
O−Ag	2.00	0.79	c d	AgO	[13] (108)
O−In	4.09	1.3	c d	InO	[13] (107)
O−Sn	5.53	1.7	c d	SnO	[13] (109)
O−Sb	5.56	1.6	c d	SbO	[13] (108)
O−Sb	3.4	1.1		Ba_2GdSbO_6 (solid)	[15]
O−Te	5.31	1.6	c d	TeO	[13] (108)
O−Te	3.76	1.2		Ba_2CaTeO_6 (solid)	[15]
O−I	5.48	1.6		IO_3^-	[4] (58)
O−Xe	6.40	1.8	c d	XeO_4	[7] (294)
O−Ba	3.97	1.8	c d	BaO	[13] (107)
O−La	5.57	2.4	c d	LaO	[13] (108)
O−Ce	6.33	2.6	c d	CeO	[13] (108)
O−Pr	5.68	2.4	c d	PrO	[13] (108)
O−Nd	3.97	1.8		$Ndac_3 \times H_2O$ (polymer)	[25] (150)
O−Eu	3.97	1.8		$Euac_3 \times 4H_2O$ (dimer) ac = acetate	[25] (150)
O−Lu	6.12	2.2	c d	LuO	[13] (108)
O−Ta	9.01	2.9	c d	TaO	[13] (109)
O−Ta	3.4	1.3		Ba_2ScTaO_6 (solid)	[15]
O−W	3.6	1.4		Ba_2CaWO_6 (solid)	[15]
O−W	9.65	3.1		WO	[13] (109)
O−Re	7.54	2.5		ReO_4^-	[4] (68)
O−Re	3.88	1.4		Ba_2NaReO_6 (solid)	[15]
O−Os	8.06	2.6	c d	OsO_4	[7] (294)
O−Pb	4.56	1.5	c d	PbO	[13] (107)
O−Bi	4.31	1.4	c d	^{209}BiO	[13] (108)
O−U	5.32	2.3	c d	UO	[13] (109)
O−U	3.22	1.5		Ba_2ZnUO_6 (solid)	[15]

LITERATURE FOR THE APPENDIX

[1] K. W. F. Kohlrausch, *Ramanspektren*, Becker & Erler, Leipzig, 1943 (reprinted by Heyden, London, 1972).

[2] G. Herzberg, Molecular Spectra and Molecular Structure, II, *Infrared and Raman Spectra of Polyatomic Molecules*, Van Nostrand, New York, 1962; pp. 201–238.

[3] Landolt-Brönstein, *Zahlenwerte und Funktionen*, Volume I, Part 2, *Molekeln I*, Springer Verlag, Berlin, 1951.

[4] H. Siebert, *Anwendungen der Schwingungsspektroskopie in der Anorganischen Chemie*, Springer Verlag, Berlin, 1966.

[5] H. J. Becher, *Fortschr. Chem. Forsch.*, 1968, **10**, 156–205.

[6] L. H. Jones, *Inorganic Vibrational Spectroscopy*, Marcel Dekker, New York, 1971.

[7] A. Fadini, *Molekülkraftkonstanten — Zur Theorie und Berechnung der Konstnaten der potentiellen Energie der Moleküle*, Steinkopff Verlag, Darmstadt, 1976.

[8] W. Sawodny, *Einige Probleme der Kraftkonstantenrechnung und -deutung*, Universität Ulm, 1976.

[9] W. Krasser, A. Fadini, E. Rozemuller, and A. J. Renouprez, *J. Mol. Struct.*, 1980, **66**, 135.

[10] A. Fadini, *Schwingungsspektroskopie — Programmsammlung für Programmrechner Texas Instruments SR*-52, Kernforschungsanlage Jülich, 1980.

[11] A. Fadini, E. Glozbach, P. Krommes, and J. Lorberth, *J. Organomet. Chem.*, 1978, **149**, 297.

[12] F.-M. Schnepel, *Schwingungsspektroskopische Untersuchungen an Substanzen mit Stickstoff-Schwefel Mehrfachbindungen*, Dissertation, University of Göttingen, 1979.

[13] K. Nakamoto, *Infrared and Raman Spectra of Inorganic and Coordination Compounds*, Wiley, New York, 1978.

[14] A. Fadini, S. Kemmler-Sack, H.-J. Schittenhelm, H.-J. Rother, and U. Treiber, *Z. Anorg. Allg. Chem.*, 1979, **454**, 49.

[15] A. Fadini, I. Jooss, S. Kemmler-Sack, G. Rauser, H.-J. Rother, and E. Schillinger, H.-J. Schittenhelm, and U. Treiber, *Z. Anorg. Allg. Chem.*, 1978, **439**, 35.

[16] S. Kemmler-Sack, I. Joos, W.-R. Cyris, and A. Fadini, *Z. Anorg. Allg. Chem.*, 1979, **453**, 153.

[17] F.-M. Schnepel, and O. Glemser, *Spectrochim. Acta*, 1981, **37A**, 257.

[18] A. Herbrechtsmeier, F.-M. Schnepel, and O. Glemser, *J. Mol. Struct.*, 1978, 50, 43.

[19] B. Schrader, *Habilitationsschrift*, University of Münster, 1968, p. 82.

[20] G. Herzberg, *Molecular Spectra and Molecular Structure*, I, *Spectra of Diatomic Molecules*, Van Nostrand, New York, 1955.

[21] H. Becher, and F. Höfler, *Spectrochim. Acta*, 1969, **25A**, 1703.

[22] W. Krasser, A. Fadini, and A. J. Renouprez, *J. Mol. Struct.*, 1980, **60**, 427.

[23] A. Fadini, H. F. Schröder, and J. Müller, *Z. Anorg. Allg. Chem.*, 1982, **488**, 121.

[24] H. Volkmann (Herausgeber), *Handbuch der Infrarot-Spektroskopie*, Verlag Chemie, Weinheim, 1972.

[25] F. Fuchs, *Darstellung, Strukturen und Eigenschaften der Lanthanoidacetate*, Dissertation, University of Tübingen, 1983.

References

[1] W. W. Coblentz, *Investigations of Infrared Spectra*, Carnegie, Washington, (1905); reprinted 1962.

[2] A. Smekal, Zur Quantentheorie der Dispersion, *Naturwissenschaften*, 1923, **43**, 873.

[3] C. V. Raman, and K. S. Krishnan, A New Type of Secondary Radiation, *Nature*, 1928, **121**, 501.

[4] W. Brügel, *Einführung in die Ultrarotspektroskopie*, Steinkopf Verlag, Darmstadt, 1969.

[5] D. A. Long, *Raman Spectroscopy*, McGraw-Hill, New York, 1977.

[6] H. Siebert, *Anwendungen der Schwingungsspektroskopie in der anorganischen Chemie*, Springer Verlag, Berlin, 1966; p. 20.

[7] K.-H. Tykto, and B. Schönfeld, *Z. Naturforsch.*, 1975, **30b**, 471.

[8] H. Günzler, H. Böck, *IR-Spectroskopie — Eine Einführung*, Verlag Chemie, Weinheim, 1975.

[9] I. Stahl, R. Mews, and O. Glemser, *Angew. Chem.*, 1980, **92**, 393.

[10] B. Schrader, and W. Meier, *DMS — Raman IR Atlas organischer Verbindungen* Verlag Chemie, Weinheim, 1974.

[11] H. Weitkamp, and R. Barth, *Infrarot-Strukturanalyse — Ein dualistisches Interpretationsschema*, Georg Thieme Verlag, Stuttgart, New York, 1972.

[12] L. J. Bellamy, *Ultrarot-Spektrum und chemische Konstitution*, Steinkopf Verlag, Darmstadt, 1966.

[13] ref. [6], p. 155.

[14] ref. [6], p. 10.

[15] M. C. Tobin, *Laser Raman Spectroscopy*, Wiley, New York, 1971.

[16] ref. [6], p. 18

[17] J. A. Koningstein, *Introduction to the Theory of the Raman Effect*, Reidel, Dordrecht, 1972; pp. 125–133.

[18] R. R. Sawyer, and F. W. Behnke, Abwasseranalyse mit Ultrarot-Spektrometern Tips 5 UR, Bodenseewerk Perkin-Elmer & Co. GmbH, Überlingen.

[19] W. H. Hohman, *J. Chem. Ed.*, 1974, **51**, 553.

[20] F. A. Cotton, *Chemical Applications of Group Theory*, Wiley, New York, 1963.

[21] ref. [6], p. 81

[22] F.-M. Schnepel, *Dissertation*, University of Göttingen, 1979.

[23] ref. [6], p. 62

[24] S. Kemmler-Sack, I. Seemann, and H.-J. Schittenhelm, *Z. Anorg. Allg. Chem.*, 1976, **422**, 115.

[25] E. Kocsárdy, and A. Heydemann, *Characterization of Kaolin Minerals of Different Origin*, 10th Int. Kaolin Symposium, Budapest, 1979.

[26] W. Kleber, *Einführung in die Kristallographie*, VEB Verlag Technik, Berlin, 1977.

[27] P. Gans, *Vibrating Molecules — An Introduction to the Interpretation of Infrared and Raman Spectra*, Chapman & Hall, London, 1971.

[28] J. Weidlein, U. Müller, and K. Dehnicke, *Schwingungsspektroskopie — Eine Einführung*, Georg Thieme Verlag, Stuttgart, New York, 1982.

[29] K. Nakamoto, *Infrared Spectra of Inorganic and Coordination Compounds*, Wiley, New York, 1970.

[30] A. Fadini, *Schwingungsspektroskopie — Programmsammlung für Programmrechner Texas Instruments SR-52*, Kernforschungsanlage Jülich, 1980.

[31] D. Steele, *Theory of Vibrational Spectroscopy*, Saunders, Philidelphia, 1971.

[32] K. Nakamoto, *Infrared and Raman Spectra of Inorganic and Coordination Compounds*, Wiley, New York, 1978; p. 278.

[33] J. Weidlein, U. Müller, and K. Dehnicke, *Schwingungsfrequenzen I Hauptgruppenelemente*, Georg Thieme Verlag, Stuttgart, New York, 1981.

[34] ref. [32], p. 136.

[35] ref. [33], p. 41.

[36] J. R. Ferraro, and J. S. Ziomek, *Introductory Group Theory and its Application to Molecular Structure*, Plenum Press, New York, 1975.

[37] ref. [32], p. 154.

[38] ref. [36], p. 116.

[39] ref. [32], p. 120.

[40] ref. [6], p. 47.

[41] ref. [31], p. 208.

[42] J. C. Decius, *J. Chem. Phys.*, 1948, **16**, 1025.

[43] S. D. Ross, *Inorganic Infrared and Raman Spectra*, McGraw-Hill, London, 1972.

[44] R. Weast (ed.), *Handbook of Chemistry and Physics*, CRC Press, Cleveland, 1974; F-202.

[45] S. J. Cyvin, *Molecular Vibrations and Mean Square Amplitudes*, Elsevier, Amsterdam, 1968; p. 118.

[46] S. Califano, *Vibrational States*, Wiley, London, 1976; pp. 90–93.

[47] ref. [32], p. 128.

[48] ref. [32], pp. 120–121.

[49] ref. [30], pp. 32–33.

[50] K. Ramaswamy, and R. Namasivayam, *Acta Phys. Pol.*, 1972, **A41**, 129.

[51] T. Wentink, *J. Chem. Phys.*, 1959, **30**, 105.

[52] A. Fadini, H. F. Schröder, and J. Müller, *Z. Anorg. Allg. Chem.*, 1982, **488**, 121.

[53] A. Fadini, E. Glozbach, P. Krommes, and J. Lorberth, *J. Organomet. Chem.*, 1978, **149**, 297.

[54] A. Fadini, *ZAMM*, 1965, **65**, T 29.

[55] A. Fadini, *Z. Naturforsch.*, 1969, **24a**, 689.

[56] A. Fadini, *Molekülkraftkonstanten — Zur Theorie und Berechnung der Konstnaten der potentiellen Energie der Moleküle*, Steinkopff Verlag, Darmstadt, 1976.

[57] A. Fadini, *J. Mol. Struct.*, 1980, **60**, 143.

[58] R. Zurmühl, *Matrizen und ihre technischen Anwendungen*, Springer Verlag, Berlin, 1961.

[59] E. B. Wilson Jr., J. C. Decius, and P. C. Cross, *Molecular Vibrations — The Theory of Infrared and Raman Vibrational Spectra*, McGraw-Hill, New York, 1955.

[60] ref. [27], pp. 36–43, 137, 139, 159–161.

[61] I. N. Bronstein, and K. A. Semendajew, *Taschenbuch der Mathematik*, Teubner Verlag, Leipzig, 1959; pp. 125–127.

[62] ref. [58], pp. 35–37.

[63] F. Matossi, *Gruppentheorie der Eigenschwingungen von Punktsystemen*, Springer Verlag, Berlin, 1961; pp. 112–123.

[64] ref. [56], pp. 23–37.

[65] N. Bjerrum, *Verh. Dtsch. Phys. Ges.* 1914, **16**, 737.

[66] K. W. F. Kohlrausch, *Ramanspektren*, Becker & Erler, Leipzig, 1943 (reprinted by Heyden, London, 1972).

[67] ref. [59], pp. 14–17, 63–76.

[68] ref. [6], pp. 5–12, 28–37, 38–87.

[69] ref. [27], pp. 4–28, 134–153.

[70] L. A. Woodward, *Introduction to the Theory of Molecular Vibrations and Vibrational Spectroscopy*, Clarendon Press, Oxford, 1972.

[71] ref. [31], pp. 70–91.

[72] ref. [46], pp. 16–31, 80–101, 136–184, 207–266.

[73] ref. [32], p. 52 (XY_2 (C_{2v})).

[74] ref. [36], pp. 111–180.

[75] ref. [56], pp. 76–81.

[76] ref. [33], p. 198.

[77] A. Fadini, and S. Kemmler-Sack, *J. Mol. Struct.*, 1980, **62**, 287 (XY_6 (O_h)).

[78] A. Fadini, S. Kemmler-Sack, H.-J. Schittenhelm, H.-J. Rother, and U. Treiber, *Z. Anorg. Allg. Chem.*, 1979, **454**, 49 (X_2Y_{12} (D_{3d})).

[79] ref. [56], pp. 149–154.

[80] L. H. Jones, *Inorganic Vibrational Spectroscopy*, Marcel Dekker, New York, 1971; p. 64.

[81] ref. [32], pp. 49–54.

[82] ref. [6], p. 50.

[83] ref. [80], p. 78.

[84] S. Pinchas, and I. Laulicht, *Infrared Spectra of Labelled Compounds*, Academic Press, London, 1971; p. 219.

[85] ref. [56], pp. 102–136.

[86] A. Fadini, *Z. Naturforsch.*, 1966, **21a**, 426.

[87] ref. [32], p. 120.

[88] ref. [56], p. 288.

[89] H. Johansen, *Z. Phys. Chem.*, 1964, **227**, 305.

[90] H. Johansen, *Z. Phys. Chem.*, 1965, **230**, 240.

[91] ref. [56], pp. 295–297.

[92] ref. [80], p. 81.

[93] ref. [6], p. 52.

[94] A. Fadini, *ZAMM*, 1964, **44**, 506.

[95] A. Fadini, *Dissertation*, TH Stuttgart, 1967.

[96] W. Sawodny, A. Fadini, and K. Ballein, *Spectrochim.*, 1965, **21**, 995.

[97] ref. [56], pp. 107–113.

[98] ref. [6], pp. 33, 47–83.

[99] G. Herzberg, *Molecular Spectra and Molecular Structure, II, Infrared and Raman Spectra of Polyatomic Molecules*, Van Nostrand, New York, 1962; pp. 201–238.

[100] ref. [56], pp. 282–283, 11–12.

[101] ref. [84], p. 202.

[102] ref. [29], p. 78.

[103] ref. [80], p. 64.

[104] ref. [30], pp. 63–66.

[105] G. E. Coates, and C. Parkin, *J. Chem. Soc.*, 1963, 421.

[106] E. Steger, I.-C. Ciurea, and A. Fadini, *Spectrochim. Acta*, 1969, **25A**, 1649.

[107] ref. [6], p. 50.

[108] ref. [6], p. 35.

[109] ref. [6], p. 40.

[110] ref. [32], p. 143.

[111] ref. [32], p. 260.

[112] W. Sawodny, *Einige Probleme der Kraftkonstantenrechnung und -deutung*, Universität Ulm, 1976.

[113] H. A. Stuart, E. Funck, and W. Müller-Warmuth, *Molekülstruktur*, Springer Verlag, Berlin, 1967; pp. 34–39.

[114] R. M. Badger, *J. Chem. Phys.*, 1934, **2**, 128.

[115] R. M. Badger, *J. Chem. Phys.*, 1935, **3**, 710.

[116] ref. [6], pp. 40, 52.

[117] ref. [80], p. 80.

[118] ref. [6], p. 52.

[119] H. Siebert, *Z. Anorg. Allg. Chem.*, 1953, **273**, 170.

[120] H. Siebert, *Z. Anorg. Allg. Chem.*, 1953, **274**, 24, 34.

[121] ref. [6], pp. 34–36.

[122] L. Pauling, *Die Natur der chemischen Bindung*, Verlag Chemie, Weinheim, 1964; p. 212.

[123] A. Fadini, and S. Kemmler-Sack, *Z. Anorg. Allg. Chem.*, 1977, **436**, 210.

[124] ref. [6], p. 47.

[125] ref. [122], p. 213.

[126] ref. [6], p. 48.

[127] ref. [6], p. 36.

[128] F.-M. Schnepel, and O. Glemser, *Spectrochim. Acta*, 1981, **37A**, 257.

[129] G. Turrell, *Infrared and Raman Spectra of Crystals*, Academic Press, London, 1972.

[130] J. C. Decius, and R. M. Hexter, *Molecular Vibrations in Crystals*, McGraw-Hill, New York, 1977.

[131] B. Schrader, *Mol. Spectroscopy*, 1978, **5**, 235.

[132] H. Pick, *Einführung in die Festkörperphysik*, Wissenschaftliche Buchgesellschaft, Darmstadt, 1978.

[133] E. J. Baran, *J. Mol. Struct.*, 1978, **48**, 441.

[134] E. J. Baran, I. L. Botto, J. C. Pedregosa, and P. J. Aymonino, *Mh. Chem.*, 1978, **109**, 41.

[135] ref. [43], pp. 95, 112–113, 115.

[136] A. Fadini, I. Jooss, S. Kemmler-Sack, G. Rauser, H.-J. Rother, E. Schillinger, H.-J. Schittenhelm, and U. Treiber, *Z. Anorg. Allg. Chem.*, 1978, **439**, 35.

[137] ref. [6], p. 60 (Hg_2Cl_2).

[138] ref. [32], p. 106 (H_2).

[139] ref. [130], pp. 284–286.

[140] J. R. Ferraro, S. S. Mitra, and C. Postmus, *Inorg. Nucl. Chem. Lett.*, 1966, **2**, 269.

[141] S. S. Mitra, C. Postmus, and J. R. Ferraro, *Phys. Rev. Lett.*, 1967, **18**, 455.

[142] itref. [129], pp. 277–285.

[143] G. Traving, *Über die Theorie der Druckverbreiterung von Spektrallinien*, Braun, Karlsruhe, 1960.

[144] ref. [36], pp. 24–31, 58–109.

[145] W. G. Fateley, F. R. Dollish, N. T. McDevitt, and F. F. Bentley, *Infrared and Raman Selection Rules for Molecular and Lattice Vibrations: The Correlation Method*, Wiley, New York, 1972.

[146] D. M. Adams, and D. C. Newton, *Tables for Factor Group and Point Group Analysis*, Beckmann, Croydon, 1970.

[147] ref. [145], pp. 171–179.

[148] ref. [130], pp. 336–341.

[149] ref. [36], pp. 225–231.

[150] H. Krishner, *Einführung in die Röntgenfeinstrukturanalyse*, Vieweg Verlag, Braunschweig, 1974.

[151] P. B. Dorrain, *Symmetrie und anorganische Strukturchemie*, Vieweg Verlag, Braunschweig, 1972; p. 7.

[152] ref. [36], pp. 29–30.

[153] S. Bhagavantam, and T. Venkatarayudu, *Theory of Group and its Application to Physical Problems*, Andhra University, Waltair, 1962.

[154] ref. [36], p. 67.

[155] ref. [129], p. 76.

[156] ref [130], pp. 10, 229–230, 238–241.

[157] D. M. Adams, and D. C. Newton, *J. Chem. Soc.* (A), 1970, 2822.

[158] N. F. M. Henry, and K. Lonsdale (eds.), *International Tables for X-Ray Crystallography I*, Kynoch Press, Birmingham, 1969; p. 274 (no. 167).

[159] ref. [36], pp. 70–109.

[160] G. Dehnicke, K. Dehnicke, H. Ahsbahs, and E. Hellner, *Ber. Bunsenges. Phys. Chem.*, 1974, **78**, 1010.

[161] A. Fadini, A. Klopsch, and B. Busch, *J. Mol. Struct.*, 1980, **64**, 201.

[162] K. Dehnicke, G. Müller, and P. Ruschke, *Z. Anorg. Allg. Chem.*, 1979, **451**, 103.

[163] B. Busch, E. Hellner, and K. Dehnicke, *Naturwissenschaften*, 1976, **63**, 531.

[164] E. Hellner, H. Ahsbahs, G. Dehnicke, and K. Dehnicke, *Naturwissenschaften*, 1974, **61**, 502.

[165] W. Klaui, and K. Dehnicke, *Chem. Ber.*, 1978, **111**, 451.

[166] H. L. Schläfer, and G. Glieman, *Einführung in die Ligandenfeldtheorie*, Akademische Verlagsgesellschaft, Frankfurt/Main, 1967.

[167] ref. [112], p. 18.

[168] G. Tatzel, *Kraftkonstantenrechnung am Beispiel von Methylverbindungen der III., IV., und V. Hauptgruppe*, Dissertation, Universität Stuttgart, 1983; pp. 91–97.

[169] A. V. Kiselev, and V. I. Lygin, *Infrared Spectra of Surface Compounds*, Wiley, New York.

[170] M. L. Hain, *Infrared Spectroscopy in Surface Chemistry*, Marcel Dekker, New York, 1967.

[171] L. H. Little, *Infrared Spectra of Absorbed Species*, Academic Press, New York, 1966.

[172] H. Ibach (ed.), *Electron Spectroscopy for Surface Analysis*, Springer Verlag, Berlin, Heidelberg, 1977.

[173] W. Krasser, H. Ervens, A. Fadini, and A. J. Renoprez, *J. Raman Spectrosc.*, 1980, **9**, 80.

[174] P. Hofman, K. Horn, R. C. Unwin, and A. M. Bradshaw, *Verh. Dtsch. Phys. Ges. (VI)*, 1980, **15**, 710.

[175] W. Krasser, A. Fadini, and A. J. Renouprez, *J. Catalysis*, 1980, **62**, 94.

[176] W. Krasser, A. Fadini, and A. J. Renouprez, *J. Mol. Struct.*, 1980, **60**, 427.

[177] W. Krasser, A. Fadini, E. Rozemuller, and A. J. Renouprez, *J. Mol. Struct.*, 1980, **66**, 135.

Index